T0211949

Lecture Notes in Computer Science 12587

More information about this subseries at http://www.springer.com/series/7412

Nadya Shusharina · Mattias P. Heinrich ·
Ruobing Huang (Eds.)

Segmentation, Classification, and Registration of Multi-modality Medical Imaging Data

MICCAI 2020 Challenges, ABCs 2020, L2R 2020, TN-SCUI 2020
Held in Conjunction with MICCAI 2020
Lima, Peru, October 4–8, 2020
Proceedings

 Springer

Editors
Nadya Shusharina (iD)
Massachusetts General Hospital
Boston, MA, USA

Mattias P. Heinrich (iD)
Universität zu Lübeck
Lübeck, Germany

Ruobing Huang (iD)
Shenzhen University
Shenzhen, China

ISSN 0302-9743   ISSN 1611-3349 (electronic)
Lecture Notes in Computer Science
ISBN 978-3-030-71826-8   ISBN 978-3-030-71827-5 (eBook)
https://doi.org/10.1007/978-3-030-71827-5

LNCS Sublibrary: SL6 – Image Processing, Computer Vision, Pattern Recognition, and Graphics

This Springer imprint is published by the registered company Springer Nature Switzerland AG
The registered company address is: Gewerbestrasse 11, 6330 Cham, Switzerland

ABCs 2020 Preface

"Anatomical Brain Barriers to Cancer Spread: Segmentation from CT and MR Images" (ABCs) was a challenge organized in conjunction with the 23rd International Conference on Medical Image Computing and Computer Assisted Intervention, MICCAI 2020 (https://abcs.mgh.harvard.edu). The goal of the challenge was to identify the best methods of segmenting brain structures that serve as barriers to the spread of brain cancers and structures to be spared from irradiation, for use in computer-assisted target definition for glioma and radiotherapy plan optimization.

In very conformal radiotherapy treatments, a higher risk of treatment failure can be attributed to inaccurately defined clinical target volume (CTV). The CTV boundary is determined by anatomical structures that are natural barriers to the spreading of tumors. For gliomas, the most common brain tumors, the boundary is defined by the falx cerebri, tentorium cerebelli, brain sinuses, and ventricles. The correct placement of the CTV boundary can spare a measurable amount of brain tissue from radiation, which will in turn reduce the risk of post-treatment neurocognitive deficit.

The quality of the treatment plan is determined not only by the precise placement of the CTV boundary but also by the accurate delineation of healthy structures that must be spared from receiving the radiation dose. These structures are routinely outlined for each treatment plan; therefore, automation will improve both efficiency of the plan creation workflow and consistency of the structure definition.

For the challenge, we provided 75 cases based on the cohort of patients diagnosed with glioblastoma and low-grade glioma who underwent radiotherapy treatment at Massachusetts General Hospital. The image data provided for training (45 cases), validation (15 cases), and testing (15 cases) consisted of the CT scans acquired for treatment planning and two diagnostic MRI scans, contrast-enhanced T_1-weighted and T_2-weighted FLAIR, of the post-operative brain. All subject image sets were manually segmented to create a set of non-overlapping structures.

The challenge was a success, with over 160 segmentation predictions submitted by the 18 participating teams. Submitted results were ranked according to the mean and standard deviation of the Dice similarity coefficient between predictions and manual segmentations. For the cerebellum, falx cerebri, brain sinuses, and tentorium cerebelli the predictions were statistically indistinguishable from the variation found in the inter-rater study for the majority of the algorithms. For the ventricles, 10 out of 18 algorithms satisfied that criterion.

We would like to thank Brian De, Kevin Diao, Soleil Hernandez, Yufei Liu, Sean Maroongroge, and Moaaz Soliman for participating in the inter-rater contouring variability study, RaySearch Laboratories for sponsoring the challenge, and all participants for their time and effort that defined the success of the challenge.

January 2021

Nadya Shusharina
Carlos Cardenas
Thomas Bortfeld

Organization

General Chair

Nadya Shusharina   Massachusetts General Hospital
and Harvard Medical School, USA

Program Committee Chairs

Carlos Cardenas   MD Anderson Cancer Center
and University of Texas, USA

Thomas Bortfeld   Massachusetts General Hospital
and Harvard Medical School, USA

Data Contributors

Moaaz Soliman   MD Anderson Cancer Center
and University of Texas, USA

Sean Maroongroge   MD Anderson Cancer Center
and University of Texas, USA

Brian De   MD Anderson Cancer Center
and University of Texas, USA

Yufei Liu   MD Anderson Cancer Center
and University of Texas, USA

Kevin Diao   MD Anderson Cancer Center
and University of Texas, USA

Soleil Hernandez   MD Anderson Cancer Center
and University of Texas, USA

L2R 2020 Preface

The Learn2Reg Challenge (Learn2Reg 2020) was a comprehensive 3D medical image registration challenge which was organized as a satellite event at the 23rd International Conference on Medical Image Computing and Computer-Assisted Intervention (MICCAI 2020). The challenge workshop was held virtually in Lima, Peru, on October 8, 2020.

Medical image registration plays a very important role in improving clinical workflows, computer-assisted interventions and diagnosis as well as for research studies involving e.g. morphological analysis. Deep learning for medical registration is currently starting to show promising advances that could improve the robustness, computation speed and accuracy of conventional algorithms to enable better practical translation. Nevertheless, there have so far not been many commonly used benchmark datasets to compare learning-based registration methods with each other and with their conventional (not trained) counterparts. To enable the development of comprehensive registration methods with similar coverage of medical imaging modalities and anatomical sites, we aimed to lower the entry barrier for new researchers to contribute to this emerging field. Motivated by the success of segmentation challenges (e.g. Medical Decathlon), we envisioned substantial improvements also in learning-based registration by providing more standardized datasets that are easily available and accessible. This also entailed a simplified challenge design that removes many of the common pitfalls for learning and applying transformations. All datasets were provided as pre-preprocessed data (resample, crop, pre-align, etc.) that can be directly employed by most conventional and learning frameworks.

Our challenge comprised four clinically relevant sub-tasks (datasets) that are complementary in nature. Participants could either individually or comprehensively address these tasks that cover both intra- and inter-patient alignment, CT, ultrasound and MRI modalities, neuro-, thorax and abdominal anatomies registration and address four of the imminent challenges of medical image registration:

learning from small datasets; estimating large deformations;
dealing with multi-modal scans; learning from limited annotations.

A total of more than 500 3D scans were made available to the public, including 22 ultrasound-MRI pairs, 30 lung CT inhale-exhale pairs, 30 inter-patient abdominal CT and 394 hippocampus MRI scans (for details please see learn2reg.grand-challenge.org/Datasets). The evaluation of the more than two dozen individual task submissions was carried out with a comprehensive evaluation pipeline, based on displacement fields, to compute the methods' performances. Since medical image registration is not limited to accurately and robustly transferring anatomical annotations, which was measured by computing target registration errors of landmarks or Dice and surface metrics of anatomical segmentations, we also incorporated a measure of transformation complexity (the standard deviation of local volume change defined by the log Jacobian determinant of the deformation) and a direct comparison of run times through either CPU or our provided Nvidia GPU backends (this was not a strict requirement for participants). All metrics were converted into significant ranks and an overall winner across the four tasks was determined in the method described in "Large Deformation Image Registration with Anatomy-aware Laplacian Pyramid Networks" by Tony Mok and Albert Chung. A perhaps surprising outcome was that conventional methods still reached the top ranks (first, second and third places) in three out of the four tasks in terms of accuracy (more results are available at https://learn2reg.grand-challenge.org/ Results/).

The proceedings of this workshop contain seven selected papers that cover a wide spectrum of conventional and learning-based registration methods and often describe novel contributions. All papers underwent a light review process by our program committee chairs.

We would like to thank all the Learn2Reg participants and co-organizers for their efforts that helped provide substantial new insights for this emerging research field and immensely contributed to the success of this challenge. We are grateful to NVIDIA and Scaleway for sponsoring the challenge.

January 2021   Mattias P. Heinrich
                                                       Alessa Hering
                                                       Lasse Hansen
                                                       Adrian Dalca

Organization

General Chair

Mattias Heinrich   University of Lübeck, Germany

Program Committee Chairs

Alessa Hering   Radboud UMC, The Netherlands,
                                     and Frauenhofer MEVIS, Germany

Lasse Hansen   University of Lübeck, Germany

Steering Committee

Adrian Dalca   A.A. Martinos Center for Biomedical Imaging
                                     MGH and MIT, USA

Data Contributors

Bram van Ginneken   Radboud UMC, The Netherlands
Yiming Xiao   Concordia University, Canada
Bennett Landman   Vanderbilt University, USA

TN-SCUI 2020 Preface

Thyroid Nodule Segmentation and Classification in Ultrasound Images Challenge (TN-SCUI 2020) is a satellite event at the 23rd International Conference on Medical Image Computing and Computer-Assisted Intervention (MICCAI 2020). The challenge TN-SCUI 2020 was held virtually in Lima, Peru, on October 4, 2020, in conjunction with the Advances in Simplifying Medical UltraSound workshop (ASMUS 2020).

The term thyroid nodule refers to an abnormal growth of thyroid cells that forms a lump within the thyroid gland. Statistical studies showed that the incidence of this disease increases with age, extending to more than 50 % of the world's population. Until recently, thyroid cancer was the most quickly increasing cancer diagnosis in the United States. It is the most common cancer in women aged 20 to 34. Although the vast majority of thyroid nodules are benign (noncancerous), a small proportion of thyroid nodules contain thyroid cancer. In order to diagnose and treat thyroid cancer at the earliest stage, it is desired to characterize nodules accurately.

Thyroid ultrasound is a key tool for thyroid nodule evaluation. It is non-invasive, real-time and radiation-free. However, it is difficult to interpret ultrasound images and recognize the subtle difference between malignant and benign nodules. The diagnosis process is thus time-consuming and heavily depends on the knowledge and the experience of clinicians.

Recently, many computer-aided diagnosis (CAD) systems have been used to alleviate this problem. However, it is usually difficult to evaluate their efficacy as no benchmark was available so far. TN-SCUI2020 aims to provide such a platform to validate all of the state-of-the-art methods and exchange new ideas.

The main topic of this TN-SCUI2020 challenge is finding automatic algorithms to accurately segment and classify thyroid nodules in ultrasound images. It provides the biggest public dataset of thyroid nodules with over 4,500 patient cases from different ages and genders, and collected using different ultrasound machines. Each ultrasound image is provided with its annotated class (benign or malignant) and a detailed delineation of the nodule. This challenge offers a unique opportunity for participants from different backgrounds (e.g. academia, industry, government, etc.) to compare their algorithms impartially.

To achieve the goals, we chose to build the challenge on the well-known platform dedicated to the biomedical imaging challenge at https://tn-scui2020.grand-challenge. org/Home/. This challenge has received a lot of attention in the related field and more than 560 teams from all over the world joined the competition. More than 1600 submissions were submitted and automatically evaluated through the platform for the final ranking based on the same test set. From the leaderboard, the top 6 teams with the highest ranks were selected as the winners of thyroid nodule segmentation and classification tasks. A rigorous review process was made by the program committee on the 6 winner solution papers for presentation on the workshop day and publication in the proceedings.

We would like to thank all the TN-SCUI 2020 organizers for their collaboration, instruction, and comments on the organizing and reviewing; we would especially like to thank the medical group for the data set collecting. We are grateful to NVIDIA and the MGI group for sponsoring the challenge. Thank you to all the participants who gave generously of their time and contributed to the success of the challenge.

December 2020

Dong Ni
Jianqiao Zhou
Alison Noble
Ruobing Huang
Tao Tan
Manh The Van

Organization

General Chair

Dong Ni   Shenzhen University, China

Program Committee Chairs

Alison Noble   University of Oxford, UK
Jianqiao Zhou   Shanghai Jiao Tong University, China
Ruobing Huang   Shenzhen University, China
Tao Tan   Eindhoven University of Technology, The Netherlands

Program Committee

Manh The Van   Shenzhen University, China
Xing Tao   Hangzhou Dianzi University, China
Rui Li   Shenzhen University, China
Xiaohong Jia   Shanghai Jiao Tong University, China
Yijie Dong   Shanghai Jiao Tong University, China

Contents

**TN-SCUI – Thyroid Nodule Segmentation and Classification
in Ultrasound Images**

ABCs – Anatomical Brain Barriers to Cancer Spread: Segmentation from CT and MR Images

Cross-Modality Brain Structures Image Segmentation for the Radiotherapy Target Definition and Plan Optimization

Nadya Shusharina[1(✉)], Thomas Bortfeld[1], Carlos Cardenas[2], Brian De[3], Kevin Diao[3], Soleil Hernandez[2], Yufei Liu[3], Sean Maroongroge[3], Jonas Söderberg[4], Moaaz Soliman[3]

[1] Division of Radiation Biophysics, Massachusetts General Hospital and Harvard Medical School, Boston, MA 02114, USA
{nshusharina,tbortfeld}@mgh.harvard.edu
[2] Department of Radiation Physics, The University of Texas, MD Anderson Cancer Center, Houston, TX 77030, USA
{CECardenas,SHernandez6}@mdanderson.org
[3] Department of Radiation Oncology, The University of Texas, MD Anderson Cancer Center, Houston, TX 77030, USA
{BSDe,KDiao,YLiu26,SMaroongroge,MSoliman}@mdanderson.org
[4] RaySearch Laboratories, Stockholm 111 34, Sweden
jonas.soderberg@raysearchlabs.com

Abstract. This paper summarizes results of the International Challenge "Anatomical Brain Barriers to Cancer Spread: Segmentation from CT and MR Images", ABCs, organized in conjunction with the MICCAI 2020 conference. Eighteen segmentation algorithms were trained on a set of 45 CT, T_1-weighted MR, and T_2-weighted FLAIR MR postoperative images of glioblastoma and low-grade glioma patients. Manual delineations were provided for the brain structures: falx cerebri, tentorium cerebelli, transverse and sagittal brain sinuses, ventricles, cerebellum (Task 1) and for the brainstem, structures of visual pathway, optic chiasm, optic nerves, and eyes, structures of auditory pathway, cochlea, and lacrimal glands (Task 2). The algorithms were tested on a set of 15 cases and received the final score for predicting segmentation on a separate 15 case image set. Multi-rater delineations with seven raters were obtained for the three cases. The results suggest that neural network based algorithms have become a successful technique of brain structure segmentation, and closely approach human performance in segmenting specific brain structures.

Keywords: Segmentation · Deep learning · Cross-modality · Radiotherapy target

1 Introduction

With evolving technology, the delivery of radiation dose to the treatment target has become very conformal, with as small as 1 mm uncertainty. Therefore, the

N. Shusharina et al. (Eds.): ABCs 2020/L2R 2020/TN-SCUI 2020, LNCS 12587, pp. 3–15, 2021.
https://doi.org/10.1007/978-3-030-71827-5_1

definition of the target itself and its consistency are becoming critical. Radiation oncologists define the target for irradiation as the gross disease which is visible on medical images, plus a margin that accounts for microscopic "invisible" spread into surrounding tissues. The margin defines the boundary of the clinical target volume (CTV). Recent studies suggested that in some very conformal treatments, higher risk of treatment failures could be attributed to inadequately or inaccurately defined CTV [1]. While the CTV cannot be defined by thresholding image intensity, its boundary can be determined by anatomical structures that are natural barriers to tumor spread. For gliomas, the most common brain tumors, the falx cerebri, tentorium cerebelli, brain sinuses, and ventricles are known to restrict the spread, and therefore these structures define the CTV boundary [2].

In current practice, CTV is manually delineated on CT image acquired for radiotherapy treatment planning to deliver a high curative radiation dose. Correct placement of the CTV boundary can spare a measurable amount of brain tissue from radiation [3], which will in turn reduce the risk of post-treatment neurocognitive deficit [4]. Identifying neuroanatomy on CT is challenging because of low soft-tissue image contrast and complex 3D shape of brain structures. This presents difficulty to follow target delineation guidelines and often results in unnecessary radiation exposure of tumor-free tissues. Undoubtedly, accurate segmentation of the barrier structures will improve definition of the CTV boundary leading to better treatment outcomes for the patients.

The quality of the treatment plan is determined not only by the precise placement of the CTV boundary but also by the accurate delineation of healthy structures that must be spared from receiving the radiation dose. The plans for glioma patients must include delineations of the brainstem, structures of visual pathway, optic chiasm, optic nerves, eyes, and lens, structures of auditory pathway, cochlea, and lacrimal glands. These structures are routinely manually outlined for each treatment plan; therefore, automation will improve both efficiency of the plan creation workflow and consistency of the structure definition.

In current clinical practice, in addition to the CT scan used for the treatment planning, each patient diagnosed with glioma and prescribed radiotherapy undergoes diagnostic post-Gadolinium T_1-weighted MR and T_2-weighted FLAIR MR imaging. Each of these modalities provides information on different types of physical properties, and therefore includes additional information relevant for image segmentation. To address the challenges of anatomy segmentation, multiple groups have proposed to combine CT images with information from MR scans [5,6]. Chartsias et al. [7] used a generative adversarial network (GAN) approach to produce synthetic MR images from CT images from a unpaired training set of CT and MR images. Synthetic images were then used for data augmentation when training a segmentation algorithm. The GAN approach was also used in [8] to develop a cross-modality deep learning segmentation algorithm, where GAN-generated pseudo MR images were used for segmentation in conjunction with real CT. In these papers, the CT and MR data were unpaired, reflecting the difficulty of obtaining matched CT and MR images for a substantial number of cases. Robust segmentation algorithms gracefully handling missing imaging modalities have been demonstrated, using an abstraction layer [9] or channel dropout during training [10].

In conjunction with the international conference on Medical Image Computing and Computer Assisted Interventions (MICCAI), we organized a challenge "Anatomical Brain Barriers to Cancer Spread: Segmentation from CT and MR Images" (ABCs). The goal of the challenge was to identify the best methods of segmenting brain structures that serve as barriers to the spread of brain cancers and structures to be spared from irradiation, for use in computer assisted target definition for glioma and radiotherapy plan optimization. For the challenge, we compiled a large image dataset acquired for patients diagnosed with glioblastoma and low-grade glioma who underwent surgery and radiotherapy treatment at Massachusetts General Hospital. Ground truth manual delineations for the anatomical brain structures were provided to developers to train their algorithms. The algorithms were optimized through testing against unseen manual delineations during the challenge and the final predictions were submitted as one-time inference on a separate, also unseen, set of cases. This paper summarizes the results of the ABCs challenge.

2   Challenge Setup

2.1   Imaging

The imaging data consisted of 75 cases of glioblastoma and low-grade glioma patients treated with surgery and adjuvant radiotherapy at Massachusetts General Hospital. The patients underwent routine post-surgical MRI examination by acquiring two MR sequences, contrast enhanced 3D-T_1 and 2D multislice-T_2 FLAIR required to define target volumes for radiotherapy treatment. CT scans were acquired after diagnostic imaging to use in radiotherapy treatment planning. Within-slice resolution (in mm) of the CT images was 0.7×0.7 (63%) 0.6×0.6 (27%) and 0.8×0.8, 0.5×0.5, 1.0×1.0 (10%) with the slice thickness of $2.5\,mm$. The resolution of MR T_1 was 1.0×1.0 (50%), 0.5×0.5 (25%), 0.9×0.9 (22%) and $1.1 \times 1.1\,mm$ (3%). The slice thickness for majority of images was $1.0\,mm$ (75%), while the rest were between 0.5 and $2.0\,mm$. For the majority of MR T_2 images, the resolution was $0.9 \times 0.9\,mm$ (74%) and varied between $0.4 \times 0.4\,mm$ to $1.0 \times 1.0\,mm$ for the rest. The slice thickness was typically $6\,mm$ (92%) although several images had the thickness of 1.0, 5.0, 6.5, and $7.0\,mm$.

2.2   Structure Labeling

The challenge was divided into two tasks; Task 1 required segmentation of the brain structures relevant for the radiotherapy target definition and Task 2 required segmentation of the structures to be included in the treatment plan optimization to minimize radiation dose to adjacent healthy organs. For the both tasks, all cases image sets were manually delineated to create a set of non-overlaying structures. For Task 1, the segmentation was done by one annotator and approved by a neuro-anatomist. For Task 2, multiple annotators, following the same annotation protocol, performed segmentation. For Task 1, ground truth labeling was obtained by manual delineation of the brain structures, falx cerebri, tentorium cerebelli, transverse and sagittal brain sinuses, ventricles, and cerebellum. As these structures are best seen on MR images (see Fig. 1), delineation was performed on the CT and T_1-weighted MR image fusion using MIM Maestro software v.6 (MIM Software Inc, Cleveland, OH, USA). We ensured that the structures adjacent to each other did not overlap and all structures were continuous without holes in each of three planes. For Task 2, manually delineated brainstem, optic chiasm, optic nerves, and eyes, lacrimal glands, and cochlea were previously delineated by certified clinical personnel and approved by the radiation oncologist for the treatment plan optimization (see Fig. 2). These structures were manually outlined on the planning CT scan which was used to create the plan.

2.3   Data Pre-processing

For each case, the two MR images were co-registered since they were acquired sequentially during the same short imaging session. The planning CT and diagnostic MR images were aligned using rigid registration with 6 degrees of freedom. All images were resampled to the resolution of 1.2 mm in each direction and cropped to the size of $164 \times 194 \times 142$ voxels. The cropped volume was centred at the point at the top of the brainstem.

2.4   Segmentation Accuracy Evaluation and Scoring

To assess the accuracy of auto-delineation we used two metrics, the volumetric Dice Similarity Coefficient (DSC) and surface Dice Similarity Coefficient (SDSC). The volumetric DSC is a voxel-wise measure of overlap of two binary image regions, it normalizes the size of overlap to the average size of the two structures:

$$\mathrm{DSC} = \frac{V_m \bigcap V_a}{(V_m + V_a)/2} \qquad (1)$$

where V_m is the set of voxels of structure m (manual ground truth) and V_a is the set of voxels of structure a (automated segmentation).

The SDSC metric assesses the distance between two surfaces relative to a given tolerance τ, see [11], providing a measure of agreement between the borders of manually and automatically segmented structures:

$$\mathrm{SDSC} = \frac{(S_m \cap B_a^\tau) + (S_a \cap B_m^\tau)}{S_m + S_a}, \qquad (2)$$

where S_m, S_a are areas of the surfaces of structures m and a, B_m^τ and B_a^τ are the border regions of thickness τ for the surfaces of structures m and a, and $S_m \cap B_a$ is the surface area of the part of S_m such that any voxel in this part is no farther than τ from S_a. SDSC ranges from 0 to 1 representing the fraction of

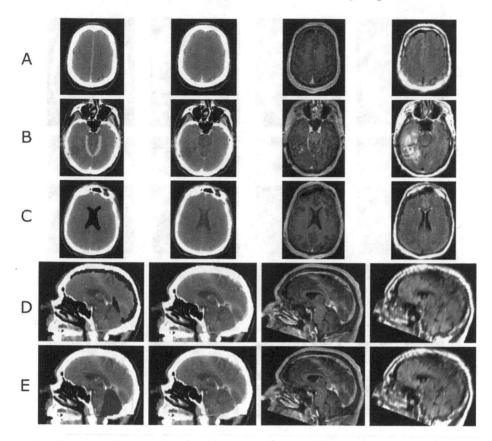

Fig. 1. Axial view (A-C) and sagittal view (D, E) of postoperative brain. On each panel from left to right: ground truth labels for Task 1, CT, MRI T_1 contrast enhanced sequence, MRI T_2 FLAIR sequence. The labels shown on the CT are: falx cerebri (A), tentorium cerebelli (B), ventricles (C), brain sinuses (D), and cerebellum (E) .

the structure border that has to be manually corrected because it deviates from the ground truth by more than the acceptable distance defined by the tolerance τ. The performance of the algorithms was evaluated using SDSC with a tolerance $\tau = 2$ mm.

For the challenge participants, the submission score was calculated as a mean of DSC and SDSC of predictions for all structures and all cases.

2.5   Organization

Participants were solicited through the MICCAI 2020 satellite events announcements. They were required to register on the challenge website (https://abcs. mgh.harvard.edu) to download manually annotated training data. After the training period of 8 weeks, the test data was released, consisting of 15 imaging sets for the participants to tune their algorithms. During testing period of

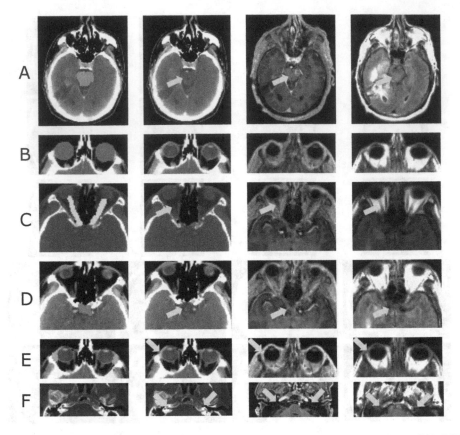

Fig. 2. On each panel from left to right: ground truth labels for Task 2, CT, MRI T_1 contrast enhanced sequence, MRI T_2 FLAIR sequence. The labels shown on CT are: brainstem (A), eyes (B), optic nerves (C), optic chiasm (D), lacrimal glands (E); cochleas (F).

three months, 18 teams submitted their predictions to be listed on the public leaderboard. To run the challenge, we created a computational platform for automated evaluation of the predictions submitted by the challenge participants. The automated Python scripted workflow consisted of calculation of the algorithm performance metrics, scoring, and ranking predictions. Two weeks prior to the MICCAI conference, the final test of 15 unseen image sets was released. Of the eighteen teams submitted the first test results, ten teams submitted their final predictions to be ranked competitively. Participants entering final ranking were asked to submit a six page summary paper describing their algorithms; of the ten submissions, five were invited to publish their full-size papers in this post-conference proceedings.

3   Results

3.1   Inter-Rater Variability Study

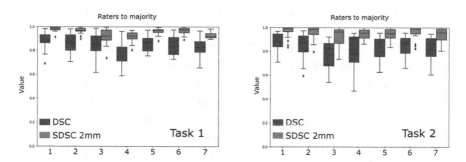

Fig. 3. Accuracy metrics, DSC and SDSC at the tolerance $\tau = 2$ mm, of inter-rater variation characterizing agreement between each of the seven raters and the majority vote contour.

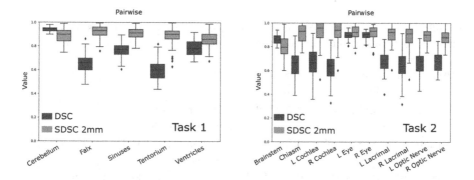

Fig. 4. Accuracy metrics, DSC and SDSC at the tolerance $\tau = 2$ mm, of inter-rater variation characterizing pairwise agreement between all raters for each structure.

In order to benchmark the accuracy of the algorithms, we conducted an inter-institutional inter-rater variability study for manual delineation of all 15 structures for three randomly selected cases. Seven raters from the Massachusetts General Hospital and MD Anderson Cancer Center were involved, five clinicians, one medical dosimetrist, and one radiographer. The raters were provided with guidelines to contour each of the 15 structures on 3 sample cases with manual contours for review prior to contouring. The seven contours per structure were collected and processed to perform the pair-wise comparison to each other and to the voxel-wise majority vote delineation using DSC and SDSC at the tolerance $\tau = 2$ mm.

Table 1. Pairwise agreement between seven raters for each structure; mean DSC and SDSC at the tolerance $\tau = 2$ mm.

Task 1	DSC	SDSC	Task 2	DSC	SDSC
Falx	0.656	0.919	Brainstem	0.860	0.801
Tentorium	0.594	0.887	Chiasm	0.653	0.914
Ventricles	0.785	0.860	Left eye	0.895	0.918
Sinuses	0.769	0.905	Right eye	0.897	0.916
Cerebellum	0.940	0.884	Left optic nerve	0.661	0.891
Mean	0.749	0.891	Right optic nerve	0.672	0.876
			Left lacrimal	0.671	0.902
			Right lacrimal	0.639	0.879
			Left cochlea	0.655	0.928
			Right cochlea	0.623	0.924
			Mean	0.723	0.895

Variability among raters was measured by the mean of DSC and SDSC between each rater and majority vote delineation for all cases and all structures (see Fig. 3 for the two Tasks). The range of DSC was (0.78, 0.881) for Task

Table 2. Mean DSC and SDSC at the tolerance $\tau = 2$ mm over 15 test cases and all structures of the algorithms scored for the competition.

Algorithm	Task 1		Task 2		Method
	DSC	SDSC	DSC	SDSC	
Jarvis	0.888	0.98	0.783	0.936	3D U-net [12], nn-Unet [13], residual blocks [14]
HILab	0.883	0.978	0.781	0.941	3D U-net [12], nn-Unet [13], coarse to fine refinement
AIViewSjtu	0.883	0.98	0.775	0.942	3D U-net [12], coarse to fine refinement [15]
MedAIR	0.88	0.977	0.786	0.935	3D U-net [12,16], nn-Unet [13], Domain Driven CNN
Sen	0.877	0.979	0.759	0.94	2D and 3D U-Net, SE-Resnext50
FREI	0.861	0.959	0.772	0.933	3D U-Net [12,16], PyTorch framework
Circle	0.858	0.979	0.736	0.927	3D UNet [12] with the C2FNAS [17]
JunMa	0.878	0.978	0.727	0.889	3D U-net [16], nn-Unet [13]
deepr	0.842	0.968	0.652	0.856	3D U-Net [12,16], PyTorch framework
SuperPod	0.802	0.935	0.229	0.266	Attention U-Net [19], V-Net [20], Inception-ResNet-v2 [21]

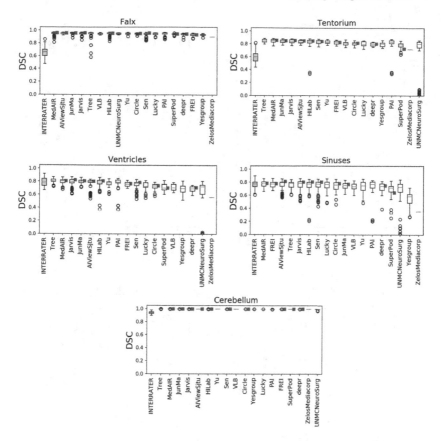

Fig. 5. Test results for the individual algorithms. Boxplots of the DSC calculated from the 15 cases of the first test set for each structure in Task 1. Red squares are mean DSC over 15 cases of the final test. Pairwise inter-rater variation is shown by cyan boxplot on the left.

1 and (0.774, 0.874) for Task 2 and the range of SDSC was (0.915, 0.979) and (0.918, 0.972) for Task 1 and 2, respectively. One of the raters provided manual delineations for the test cases used by the challenge participants to make their predictions. The rater's mean accuracy metrics (DSC and SDSC) were within the ranges: 0.828 and 0.927 for Task 1, and 0.812 and 0.94 for Task 2.

The accuracy metrics obtained by pairwise analysis comparing the mean agreement among all raters for all cases for each individual structure (see Fig. 4 for the two tasks) provided a baseline for an automated segmentation algorithm performance.

3.2   Performance of the Algorithms

Over 160 segmentation predictions were collected from the 18 participating teams during the three month algorithm optimization period. The submitted

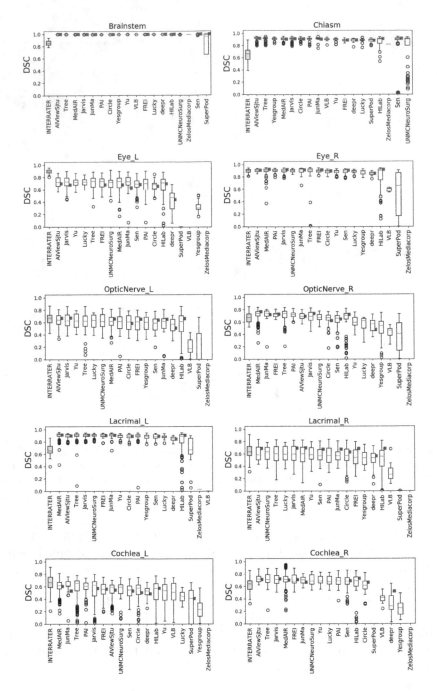

Fig. 6. Test results for the individual algorithms. Boxplots of the DSC calculated from the 15 cases of the first test set for each structure in Task 2. Red squares are mean DSC over 15 cases of the final test. Pairwise inter-rater variation is shown by cyan boxplot on the left.

results have been compared according to the mean of the DSC and SDSC between predictions and manual delineations. The results are compiled in Table 2 and presented in Figs. 5 and 6 where the results of the final test are also shown.

4   Summary and Conclusion

In this paper we presented the results of the ABCs challenge organized in conjunction with the MICCAI 2020 conference that was setup to identify the best algorithms for automated segmentation of brain structures that are used to define the clinical target volume (CTV) for radiotherapy treatment of glioma patients and the structures used for the treatment plan optimization. Accurate placement of the CTV boundary is the key to defining success of the treatment. For highly conformal dose delivery techniques, the treatment setup uncertainties are as low as 1.7 mm and do not exceed 3.5 mm [22]. The uncertainty of defining CTV boundary should not be larger than the targeting uncertainties. Therefore, clinically acceptable accuracy of auto-segmentation, defined by the SDSC, should be at least 0.95 for the tolerance $\tau = 2$ mm. The challenge showed that deep learning algorithms with multi-modality inputs can generate high quality segmentations. Specifically, four of five brain structures, tentorium cerebelli (0.95), brain sinuses (0.96), ventricles (0.96), and cerebellum (0.98) were segmented with the clinically acceptable accuracy. For the falx cerebri (0.93), the accuracy is very close to that definition.

As the degree of automation in routine clinical practice increases, rigorous evaluation of algorithm performance and extensive discussions between developers, clinicians, and regulatory authorities will become more important [23]. Although specific approval criteria are still under development, it is likely that they will include algorithmic segmentations being statistically indistinguishable from human expert ones. In the ABCs challenge, algorithms for Task 1 demonstrated a remarkable consistency of DSC for the final test (Fig. 5). For ventricles and sinuses, the leading algorithms showed the mean and variance of the DSC indistinguishable from those of seven human experts. For the falx cerebri, tentorium cerebelli, and cerebellum, algorithms were extremely consistent, unlike the human experts. As the ground truth segmentation for Task 1 structures were performed by a single expert, one cannot exclude that the algorithms could have learned and imitated this person's approach to segmentation. However, a similar pattern was also observed in Task 2 (Fig. 6), where ground truth segmentations were made by several independent experts (different from those in the inter-rater study) in routine clinical practice. Automated segmentations of the brainstem and optic chiasm were extremely consistent and showed much lower variance than the seven human experts. For most of the other structures, the average DSC over the 15 final test cases (red squares) fell in the range of the variance between the human experts (cyan boxes). That said, further work is needed to quantify the variation in human expert segmentations for a larger number of cases and a broader representations of experts from different institutions. Taken together, the results of the ABCs challenge suggest that neural network based algorithms have become a successful technique of brain structure segmentation, and closely approach human performance in segmenting specific brain structures.

References

1. Engels, B., Soete, G., Verellen, D., Storme, G.: Conformal arc radiotherapy for prostate cancer: increased biochemical failure in patients with distended rectum on the planning computed tomogram despite image guidance by implanted markers. Int. J. Radiat. Oncol. Biol. Phys. **74**(2), 388–391 (2009)
2. Niyazi, M., Brada, M., Chalmers, A.J., Combs, S.E., Erridge, S.C., Fiorentino, A., et al.: ESTRO-ACROP guideline "target delineation of glioblastomas". Radiother. Oncol. **118**(1), 35–42 (2016)
3. Kruser, T.J., et al.: NRG brain tumor specialists consensus guidelines for glioblastoma contouring. J. Neuro-Oncol. **143**(1), 157–166 (2019). https://doi.org/10.1007/s11060-019-03152-9
4. Karunamuni, R., et al.: Dose-dependent cortical thinning after partial brain irradiation in high-grade glioma. Int. J. Rad. Oncol. Biol. Phys. **94**(2), 297–304 (2016)
5. Valindria, V.V., et al.: Multi-modal learning from unpaired images: application to multi-organ segmentation in CT and MRI. In: 2018 IEEE Winter Conference on Applications of Computer Vision (WACV), Lake Tahoe, NV, pp. 547–556 (2018)
6. Yang, J., Dvornek, N.C., Zhang, F., Chapiro, J., Lin, M.D., Duncan, J.S.: Unsupervised domain adaptation via disentangled representations: application to cross-modality liver segmentation. In: Shen, D., et al. (eds.) MICCAI 2019. LNCS, vol. 11765, pp. 255–263. Springer, Cham (2019). https://doi.org/10.1007/978-3-030-32245-8_29
7. Chartsias, A., Joyce, T., Dharmakumar, R., Tsaftaris, S.A.: Adversarial image synthesis for unpaired multi-modal cardiac data. In: Tsaftaris, S.A., Gooya, A., Frangi, A.F., Prince, J.L. (eds.) SASHIMI 2017. LNCS, vol. 10557, pp. 3–13. Springer, Cham (2017). https://doi.org/10.1007/978-3-319-68127-6_1
8. Jue, J., et al.: Integrating cross-modality hallucinated MRI with CT to aid mediastinal lung tumor segmentation. In: Shen, D., et al. (eds.) MICCAI 2019. LNCS, vol. 11769, pp. 221–229. Springer, Cham (2019). https://doi.org/10.1007/978-3-030-32226-7_25
9. Havaei, M., Guizard, N., Chapados, N., Bengio, Y.: HeMIS: hetero-modal image segmentation. In: Ourselin, S., Joskowicz, L., Sabuncu, M.R., Unal, G., Wells, W. (eds.) MICCAI 2016. LNCS, vol. 9901, pp. 469–477. Springer, Cham (2016). https://doi.org/10.1007/978-3-319-46723-8_54
10. Shen, Y., Gao, M.: Brain tumor segmentation on MRI with missing modalities. In: Chung, A.C.S., Gee, J.C., Yushkevich, P.A., Bao, S. (eds.) IPMI 2019. LNCS, vol. 11492, pp. 417–428. Springer, Cham (2019). https://doi.org/10.1007/978-3-030-20351-1_32
11. Nikolov, S., Blackwell, S., Mendes, R., De Fauw, J., Meyer, C., Hughes, C., et al.: Deep learning to achieve clinically applicable segmentation of head and neck anatomy for radiotherapy. e-prints [Internet]. arXiv:1809.04430 (2018)
12. Ronneberger, O., Fischer, P., Brox, T.: U-Net: convolutional networks for biomedical image segmentation. In: Navab, N., Hornegger, J., Wells, W.M., Frangi, A.F. (eds.) MICCAI 2015. LNCS, vol. 9351, pp. 234–241. Springer, Cham (2015). https://doi.org/10.1007/978-3-319-24574-4_28

13. Isensee, F., et al.: nnu-net: self-adapting frame- work for U-Net-based medical image segmentation. arXiv preprint arXiv:1809.10486 (2018)
14. He, K., Zhang, X., Ren, S., Sun, J.: Deep residual learning for image recognition. In: Proceedings of the IEEE Conference on Computer Vision and Pattern Recognition, pp. 770–778 (2016)
15. Chen, H., Wang, X., Huang, Y., Wu, X., Yu, Y., Wang, L.: Harnessing 2D networks and 3D features for automated pancreas segmentation from volumetric CT images. MICCAI **6**, 339–347 (2019)
16. Cicek, O., Abdulkadir, A., Lienkamp, S.S., Brox, T., Ronneberger, O.: 3D U-Net: learning dense volumetric segmentation from sparse annotation. arXiv e-prints [Internet]. (2016)
17. Yu, Q., et al.: C2FNAS: coarse-to-fine neural architecture search for 3D medical image segmentation. In: Proceedings of the IEEE/CVF Conference on Computer Vision and Pattern Recognition (2020)
18. Clevert, D.-A., Unterthiner, T., Hochreiter, S.: Fast and accurate deep network learning by exponential linear units (ELUs). arXiv e-prints [Internet]. arXiv:1511.07289 (2016)
19. Oktay, O., et al.: Attention U-Net: learning where to look for the pancreas. arXiv preprint arXiv:1804.03999 (2018)
20. Milletari, F., Navab, N., Ahmadi, S.A.: V-Net: fully convolutional neural networks for volumetric medical image segmentation. In: Proceedings - 2016 4th International Conference on 3D Vision, 3DV 2016, pp. 565–571 (2016). https://doi.org/10.1109/3DV.2016.79
21. Szegedy, C., Ioffe, S., Vanhoucke, V., Alemi, A.A.: Inception-v4, inception-ResNet and the impact of residual connections on learning. In: 31st AAAI Conference Artificial Intelligence AAAI 2017, pp. 4278–4284 (2017)
22. Oh, S.A., Yea, J.W., Kang, M.K., Park, J.W., Kim, S.K.: Analysis of the setup uncertainty and margin of the daily ExacTrac 6D image guide system for patients with brain tumors. PLoS ONE **11**(3), e0151709 (2016)
23. Cardenas, C.E., Yang, J., Anderson, B.M., Court, L.E., Brock, K.B.: Advances in auto-segmentation. Semin. Radiat. Oncol. **29**(3), 185–197 (2019)

Domain Knowledge Driven Multi-modal Segmentation of Anatomical Brain Barriers to Cancer Spread

Xiaoyang Zou[1(✉)] and Qi Dou[2]

[1] School of Engineering Sciences, Huazhong University of Science and Technology, Wuhan, China
xyzou@hust.edu.cn
[2] Department of Computer Science and Engineering, The Chinese University of Hong Kong, Hong Kong, China

Abstract. It is important to accurately segment anatomical brain barriers to cancer spread with multi-modal images, in order to assist definition of the clinical target volume (CTV). In this work, we explore a multi-modal segmentation method largely driven by domain knowledge. We apply 3D U-Net as the backbone model. In order to reduce the learning difficulty of deep convolutional neural networks, we employ a label merging strategy for the symmetrical structures which have both left and right labels, to highlight the structural information regardless of the locations. Moreover, considering the existence of visual preference for certain modality and mismatches in co-registration, we adopt a multi-modality ensemble strategy for multi-modal learning to enable the models better driven by domain knowledge of this task, which is different from fully data-driven methods, like early fusion strategy for multi-modal images. By contrast, multi-modality ensemble strategy yields better segmentation results. Our method achieved an average score of 0.895 on MICCAI 2020 Anatomical Brain Barriers to Cancer Spread Challenge's final test dataset (https://abcs.mgh.harvard.edu/.). Detailed methodologies and results are described in this technical report (This work was done when X. Zou did remote internship with CUHK.).

1 Introduction

Glioma is the most common type of malignant brain tumors with an incidence of 6.9 per 100,000 population [8], comprising about 25.5% of all brain and other central nervous system tumors and 80.8% of all malignant brain tumors [10]. An effective treatment for gliomas is radiotherapy [7]. In the process of tumor's radiotherapy planning, accurate definition of the clinical target volume (CTV) is a highly critical step. Inadequate definition of CTV may lead to geometric miss of the tumors, resulting in underdosage of certain areas in radiotherapy, which may lead to higher risk of recurrence for the patients [15]. Actually, brain anatomical structures can serve as a natural barrier to the spreading of brain tumors. Thus,

© Springer Nature Switzerland AG 2021
N. Shusharina et al. (Eds.): ABCs 2020/L2R 2020/TN-SCUI 2020, LNCS 12587, pp. 16–26, 2021.
https://doi.org/10.1007/978-3-030-71827-5_2

their boundaries can effectively assist the definition of CTV [14]. In addition, in order to prevent the sensitive and important healthy organs from being affected by radiation during radiotherapy, some healthy organs also need to be accurately delineated (like eyes, cochleae, optic nerves, chiasm, etc.) [9]. However, manual delineation of brain anatomical structures and healthy organs is time-consuming and error-prone. Thus, developing a fully automatic segmentation algorithm will effectively improve the efficiency and consistency of radiotherapy planning [3], which is exactly what the Anatomical Brain Barriers to Cancer Spread (ABCs) Challenge is aimed at.

To conduct this task, multi-modal images are often required in clinical practice. Specifically, Computed Tomography (CT) and multi-sequence Magnetic Resonance Imaging (MRI) can visualize different features of brain anatomical structures [13]. There exist both shared information and complementary information between different modalities. Looking for a multi-modal learning strategy to exploit valuable modality information may improve the segmentation performance effectively.

In recent years, deep convolutional neural networks (DCNNs) have become the de facto method for automated medical image segmentation. Ronneberger et al. proposed U-Net, a landmark network architecture for medical image segmentation [12]. Çiçek et al. proposed 3D U-Net for 3D medical image segmentation on this basis [1]. Isensee et al. completed the nnU-Net training framework, making U-Net achieve a relatively high level [4–6]. On this basis, two strategies are often used for the fusion of co-registered multi-modal images, which are early fusion and late fusion [2]. The early fusion is to concatenate multi-modal images as different channels of the network's input, which has already been widely used for brain tissue's segmentation in multi-sequence MR images [17]. And the late fusion is to fuse multi-modal images at a semantic level in the middle of the network. However, most existing methods are merely data-driven, without carefully consideration to incorporate valuable domain knowledge into learning process.

In this work, instead of creating some novel DCNN architectures, we primarily focus on exploring how to facilitate the model learning with domain knowledge of this task. We employ the label merging strategy for symmetrical structures, which have both left and right labels, to reduce the learning difficulty of DCNNs. Thus, structural information can be highlighted for the network to learn. Besides, we adopt multi-modality ensemble strategy to help the network learn the domain knowledge embedded in multi-modal images, thus to realize a domain knowledge driven model instead of a fully data-driven model. We find that, with the existence of visual preference for certain modality and mismatches in co-registration, the segmentation performance of using multi-modality ensemble strategy goes beyond that of using early fusion strategy for multi-modal images.

2   Methods

In this section, we present our proposed methods including label merging strategy for symmetric structures and multi-modality ensemble strategy. In addition, we show details about the dataset, network architecture and training protocol.

2.1   Data Description

In this work, we use the 45 cases training data which are officially provided by ABCs challenge for training, and do not use any external data. In addition to this, 15 cases are used for validation leaderboard and another 15 cases are used for final test scoring. Each case contains images of three modalities, including CT, T1-weighted MR and T2-weighted FLAIR MR images. And the multi-modal images are aligned in same size and resolution by co-registration and resampling. What is noteworthy is that, in the final submission, we only use CT and T1-weighted MR images for training, without any T2-weighted FLAIR MR images.

2.2   Network Architecture

All models are trained with plain 3D U-Net implemented on nnU-Net training framework[1], as shown in Fig. 1. The network consists of two paths, which are down-sampling path and up-sampling path. In each path, there are 4 stages of down-sampling or up-sampling. And each stage consists of two blocks. In down-sampling path, the first block includes a $3 \times 3 \times 3$ convolutional layer strided $1 \times 2 \times 2$ or $2 \times 2 \times 2$ for images down-sampling, an instance normalization layer and then a leaky ReLU layer. The second block includes a $3 \times 3 \times 3$ convolutional layer strided $1 \times 1 \times 1$ with padding, an instance normalization layer and then a leaky ReLU layer. In up-sampling path, we just replace the convolutional layer in first block with transposed convolutional layer. The final block consists of a $1 \times 1 \times 1$ convolutional layer and a softmax layer. Skip connections are used to forward the information from down-sampling path to up-sampling path.

2.3   Label Merging for Symmetric Structures

The brain anatomical structures often have typical sagittal plane's symmetry. And some of them can be directly divided into left and right two parts, such as eyes, optic nerves, cochleae, lacrimal glands, etc. In Task 2 of ABCs Challenge, the left and right parts of these structures are distinguished with two labels, which are scored independently. Therefore, it's of great significance to identify the structures' left and right parts correctly.

In a sense, the difference in location between left and right can be regarded as a kind of location information, which is relatively independent of structural information. In fact, this kind of location information does not change case by

---

[1] https://github.com/MIC-DKFZ/nnUNet.

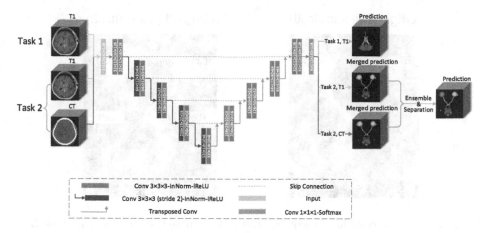

Fig. 1. The plain 3D U-Net architecture and general training pipeline that we used in ABCs challenge. Note that N in the final block is the number of label categories within the background.

case, especially when the images are scanned in the same direction. For DCNNs, such kind of redundant location information means a great learning difficulty, and may drown out the structural information. Actually, it is not necessary for DCNNs to learn such fixed location information. Instead, judgments can be made directly from left to right based on the predictions during post-processing.

In order to help DCNNs get rid of the difficulty of learning location information, we firstly merge the left and right labels of structures in Task 2 (see Fig. 2). In this way, the left and right structures have the same label. The DCNNs no longer need to learn the difference between left and right structures, but only focus on structural information.

Our model is trained with the new merged ground truth. However, predictions based on this model do not contain any location information. Thus, during post-processing, the left and right parts need to be separated again according to their actual locations in the predictions. In this challenge, images are pre-processed to same direction and have good symmetry. Based on this, we find that the y-z plane can be regarded as the sagittal plane, so as to determine whether the structure is located on the left or right side. For example, we can simply consider the minimum to half of the x-axis as the right part and the maximum to half of the x-axis as the left part. It has been verified that such a judgment is completely right for this data set.

In this way, the evaluation metrics will not decline due to the wrong judgment of left and right of DCNNs, and the segmentation task can be focused on segmenting the structures themselves.

Original groundtruth   Merged groundtruth

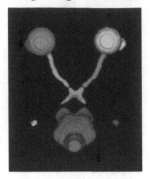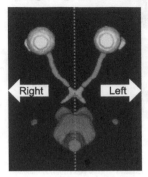

Fig. 2. An example of label merging strategy for symmetric structures in Task 2.

2.4 Multi-modality Ensemble

By observing the characteristics of the data set, we find that visual preference for certain modal images always exists when segmenting different structures or dealing with different tasks. For example, Fig. 3 shows the brainstem in different modal images. We can see that the brainstem in T1-weighted MR images has the highest contrast and the best visual preference, compared with that in CT images or T2-weighted FLAIR MR images. In other words, for the brainstem, visual preference is for T1-weighted MR images.

What's more, there exists obvious mismatches in the co-registration of multi-modal images. Considering that each modality shares the ground truth, when we compare each single modal image with the ground truth, it is not difficult to discover that some structures are mismatched with the ground truth. As Fig. 4 shows, we compare the same slice from all modal images and the ground truth in both Task 1 and Task 2. The mismatches between the location of actual structures and the ground truth are highlighted, which mainly exist at ventricles, sinuses, eyes, etc. And we can see that T1-weighted MR images almost have no mismatch in Task 1, while CT images almost have no mismatch in Task 2.

Although early fusion is a commonly used multi-modal fusion strategy, DCNNs are hard to distinguish the domain difference mentioned above with data-driven method only. It is necessary to try a domain knowledge driven method of this task. Thus, we further explore the segmentation performance when training with each single modality, hoping to train a better domain knowledge driven model with multi-modality ensemble strategy. Finally, we use T1-weighted MR images for training to segment all the structures in Task 1 and the brainstem in Task 2, while use CT images for training to segment all the structures in Task 2 except the brainstem.

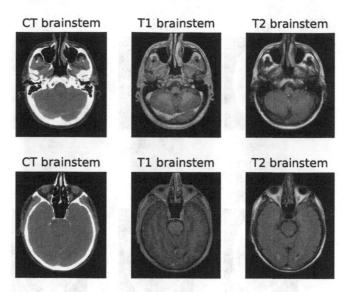

Fig. 3. The brainstem is shown in three different modal images. The CT value of the CT images have been clipped to −100–300 in order to improve the contrast.

2.5   Training Protocol

Pre-processing methods include non-zero region cropping and normalization. For CT images, we clip the CT value from 0.5% to 99.5% for training to reduce less informative pixels. All CT cases are normalized together by z score, considering that CT value is quantitative. For MR images, each case is individually normalized by z score due to the distribution difference. Patches are extracted randomly with a size of [112, 160, 128]. And the batch size is 2. Images are augmented through random rotation, scaling, mirroring, gamma, brightness and contrast transformation. Stochastic gradient descent optimizer is used for training, with an initial learning rate of 1e-2 and a weighted decay of 3e-5. The total loss is the sum of dice loss and cross-entropy loss. All models are trained using only one NVIDIA TITAN Xp GPU and PyTorch library [11].

In the final submission, we adopt the model ensemble strategy to improve the model's robustness. Models are test on the validation leaderboard every 100 epochs trained. We finally select the best two models to employ model ensemble strategy. The final prediction is obtained by averaging the softmax output of the two models and taking 0.5 as the threshold for binarization.

3   Results

Both of dice similarity coefficient (DSC) and surface dice similarity coefficient (SDSC) at the tolerance of 2 mm are used to evaluate the accuracy of our automatic delineation. Specific calculations are based on an open source library from

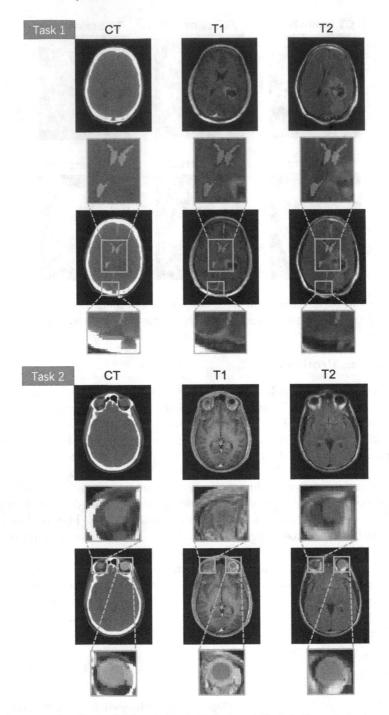

Fig. 4. An example of mismatches in both Task 1 and Task 2. In each task, same slice is taken from all modal images and the ground truth. The contrast is auto-adjusted by ITK-SNAP [16].

DeepMind[2], which is recommended by the challenge organizers. And the average score over all test data of both DSC and SDSC is taken as the final evaluation. All experiments were completed during the ABCs 2020 Challenge, and all models were evaluated by 5-fold cross-validation on the training set or with data for the validation leaderboard.

Table 1 shows the effectiveness of label merging strategy for symmetric structures. For Task2, we compare the average DSC with and without label merging strategy and calculate their difference, under the condition of training with the same plain 3D U-Net and T1-weighted MR images. It can be seen that label merging strategy realizes a huge improvement of DSC in Task 2, about 33.3%, especially to the structures that can be divided into left and right two parts.

Table 1. Models are only trained on T1-weighted MR images. The DSC scores trained based on the original ground truth are shown in column T1, and the DSC scores trained based on the ground truth after label merging are shown in column T1+Label merging. The column Difference calculates the difference value between those two DSC scores. The results are evaluated based on 5-fold cross-validation.

Task 2 structure	T1	T1 + label merging	Difference
Brainstem	0.885	0.899	+0.014
Chiasm	0.450	0.510	+0.060
Left Cochlea	0.340	0.715	+0.375
Right Cochlea	0.464	0.694	+0.230
Left Eye	0.671	0.896	+0.225
Right Eye	0.692	0.910	+0.218
Left Lacrimal	0.504	0.672	+0.168
Right Lacrimal	0.527	0.676	+0.149
Left OpticNerve	0.474	0.673	+0.199
Right OpticNerve	0.493	0.683	+0.190
Average	**0.550**	**0.733**	**+0.183**

On the basis of adopting label merging strategy, we compare the results of the early fusion strategy of multi-modal images with the multi-modality ensemble strategy, as shown in Table 2. The scores listed here are based on the validation set for leaderboard. It can be seen that the results based on our multi-modality ensemble strategy completely exceed the results based on early fusion strategy. Besides, segmenting brainstem with T1-weighted MR images can help achieve better results indeed.

Through model ensemble, we further improve our performance on the validation leaderboard. Table 3 shows the final results of top 5 teams on the ABCs 2020 Challenge's validation leaderboard. Our method surpassed all the other 20

---

[2] https://github.com/deepmind/surface-distance.

Table 2. Comparison of early fusion strategy and multi-modality ensemble strategy. The early fusion adopted here is to concatenate all the three modal images as the networks input. The contents in parentheses indicate the scope of different modal images.

Method	Task 1 DSC	Task 1 SDSC	Task 2 DSC	Task 2 SDSC
Task 1: early fusion Task 2: early fusion	0.867	0.973	0.755	0.921
Task 1: T1 (all) Task 2: CT (all)	0.877	0.975	0.777	0.924
Task 1: T1 (all) Task 2: T1 (brainstem), CT (others)	**0.877**	**0.975**	**0.779**	**0.929**

competing teams and ranked the 1st place on the validation leaderboard by a narrow margin. And we still won the 3rd place for the final test scoring with an average score of 0.895, which is 0.002 behind the 1st place.

Table 3. Top 5 results on ABCs 2020 Challenge's validation leaderboard. Submissions are ranked by the overall average score of DSC and SDSC of each structure in Task1 and Task2. Note that the data used for validation leaderboard and the data used for final scoring are not same batch data. So this result does not represent the final scoring of the challenge.

Team name	Task 1 DSC	Task 1 SDSC	Task 2 DSC	Task 2 SDSC	Average score
MedAIR (ours)	**0.877**	**0.975**	0.781	0.930	**0.891**
Holo	0.876	0.974	0.782	0.929	0.890
Tree	0.875	0.974	0.782	0.929	0.890
AIViewSjtu	0.869	0.974	0.772	0.939	0.888
HILab	0.869	0.972	0.771	0.933	0.886
Sen	0.863	0.972	0.758	0.930	0.881

4   Discussion

It should be noted that the method we use to separate the merged labels in predictions has some limitations. We just simply make the judgment between left and right according to the x-coordinate, considering that all images provided by ABCs 2020 Challenge have been pre-processed to same direction and have good symmetry. However, for images with different direction and poor symmetry, a more essential approach to solve this problem is required. For example, we may need to convert the images to the same direction, or even find and combine the middle surface place of the symmetric structures for judgment.

For the segmentation of multi-modal images, we believe that different modal images contribute differently to the segmentation of different structures or different tasks. Although we have discovered that multi-modality ensemble strategy shows better segmentation performance compared with early fusion of multi-modal images in ABCs challenge, the selection of best modality needs further study. Our final selection suggests two possibilities. The best modality might be the one has the best visual contrast or the one that is used for the data set's labeling. In addition, considering the mismatches of co-registration, we need to think about whether it is worthwhile to adopt a multi-modality fusion strategy if obvious co-registration noise exists.

What's more, with the help of the powerful nnU-Net training framework, it seems using some novel DCNN architectures or training methods cannot make significant improvements. We have tried to use 3D U-Net with residual blocks, and to modify the loss function as the sum of dice loss and focal loss or the sum of dice loss and weighted cross entropy loss. However, we did not obtain any improvement on the validation leaderboard. Compared with this, it seems more important to make the model become more robust. Thus, we adopt the model ensemble strategy here, which is to average the softmax outputs of the top two best performing models that we submitted on the leaderboard and then predict with a 0.5 threshold value. This strategy helps to improve the robustness of the model, thus enabling the model to achieve better performance.

5 Conclusion

In this work, we use plain 3D U-Net to segment anatomical brain barriers to cancer spread. For Task 2, we adopt a label merging strategy for symmetrical structures to reduce the network's difficulty of learning location information. That is, the DCNNs only need to focus on structural information at the semantic level, then the left and right parts will be automatically separated according to their real location in the predictions. This strategy can significantly improve the DCNNs' segmentation performance on those structures that have both left and right labels. In addition, we adopt multi-modality ensemble strategy to facilitate the model better driven by domain knowledge. The domain knowledge is important for the networks to learn especially when visual preference for certain modality and mismatches in co-registration exist. And the segmentation performance surpasses that of employing only data-driven method like multi-modal early fusion strategy.

References

1. Çiçek, Ö., Abdulkadir, A., Lienkamp, S.S., Brox, T., Ronneberger, O.: 3D U-Net: learning dense volumetric segmentation from sparse annotation. In: Ourselin, S., Joskowicz, L., Sabuncu, M.R., Unal, G., Wells, W. (eds.) MICCAI 2016. LNCS, vol. 9901, pp. 424–432. Springer, Cham (2016). https://doi.org/10.1007/978-3-319-46723-8_49
2. Dou, Q., Liu, Q., Heng, P., Glocker, B.: Unpaired multi-modal segmentation via knowledge distillation. IEEE Trans. Med. Imaging **39**, 2415–2425 (2020)

3. Duanmu, H., et al.: Automatic brain organ segmentation with 3D fully convolutional neural network for radiation therapy treatment planning. In: 2020 IEEE 17th International Symposium on Biomedical Imaging (ISBI), pp. 758–762. IEEE (2020)

4. Isensee, F., Jäger, P.F., Kohl, S.A., Petersen, J., Maier-Hein, K.H.: Automated design of deep learning methods for biomedical image segmentation. arXiv preprint arXiv:1904.08128 (2019)

5. Isensee, F., et al.: nnU-Net: self-adapting framework for u-net-based medical image segmentation. arXiv preprint arXiv:1809.10486 (2018)

6. Isensee, F., Petersen, J., Kohl, S.A., Jäger, P.F., Maier-Hein, K.H.: nnU-Net: breaking the spell on successful medical image segmentation, vol. 1, pp. 1–8. arXiv preprint arXiv:1904.08128 (2019)

7. Laperriere, N., Zuraw, L., Cairncross, G., Cancer Care Ontario Practice Guidelines Initiative Neuro-Oncology Disease Site Group, et al.: Radiotherapy for newly diagnosed malignant glioma in adults: a systematic review. Radiother. Oncol. **64**(3), 259–273 (2002)

8. Li, K., et al.: Trends and patterns of incidence of diffuse glioma in adults in the united states, 1973–2014. Cancer Med. **7**(10), 5281–5290 (2018)

9. Mlynarski, P., Delingette, H., Alghamdi, H., Bondiau, P.Y., Ayache, N.: Anatomically consistent segmentation of organs at risk in MRI with convolutional neural networks. arXiv preprint arXiv:1907.02003 (2019)

10. Ostrom, Q.T., et al.: CBTRUS statistical report: primary brain and other central nervous system tumors diagnosed in the united states in 2012–2016. Neuro-Oncol. **21**(Suppl._5), v1–v100 (2019)

11. Paszke, A., et al.: PyTorch: an imperative style, high-performance deep learning library. In: Advances in Neural Information Processing Systems, pp. 8026–8037 (2019)

12. Ronneberger, O., Fischer, P., Brox, T.: U-Net: convolutional networks for biomedical image segmentation. In: Navab, N., Hornegger, J., Wells, W.M., Frangi, A.F. (eds.) MICCAI 2015. LNCS, vol. 9351, pp. 234–241. Springer, Cham (2015). https://doi.org/10.1007/978-3-319-24574-4_28

13. Rumboldt, Z., Castillo, M., Huang, B., Rossi, A.: Brain Imaging with MRI and CT: An Image Pattern Approach. Cambridge University Press, Cambridge (2012)

14. Shusharina, N., Söderberg, J., Edmunds, D., Löfman, F., Shih, H., Bortfeld, T.: Automated delineation of the clinical target volume using anatomically constrained 3D expansion of the gross tumor volume. Radiother. Oncol. **146**, 37–43 (2020)

15. Weiss, E., Hess, C.F.: The impact of gross tumor volume (GTV) and clinical target volume (CTV) definition on the total accuracy in radiotherapy. Strahlentherapie und Onkologie **179**(1), 21–30 (2003)

16. Yushkevich, P.A., Gao, Y., Gerig, G.: ITK-SNAP: an interactive tool for semi-automatic segmentation of multi-modality biomedical images. In: 2016 38th Annual International Conference of the IEEE Engineering in Medicine and Biology Society (EMBC), pp. 3342–3345. IEEE (2016)

17. Zhang, W., et al.: Deep convolutional neural networks for multi-modality isointense infant brain image segmentation. NeuroImage **108**, 214–224 (2015)

Ensembled ResUnet for Anatomical Brain Barriers Segmentation

Munan Ning[✉], Cheng Bian, Chenglang Yuan, Kai Ma, and Yefeng Zheng

Tencent Jarvis Lab, Shenzhen, China
{masonning,tronbian,nikoyuan,kylekma,yefengzheng}@tencent.com

Abstract. Accuracy segmentation of brain structures could be helpful for glioma and radiotherapy planning. However, due to the visual and anatomical differences between different modalities, the accurate segmentation of brain structures becomes challenging. To address this problem, we first construct a residual block based U-shape network with a deep encoder and shallow decoder, which can trade off the framework performance and efficiency. Then, we introduce the Tversky loss to address the issue of the class imbalance between different foreground and the background classes. Finally, a model ensemble strategy is utilized to remove outliers and further boost performance.

1 Introduction

Gliomas are the most common type of brain tumors, originating from glial cells in the human brain. In clinical radiation therapy, the precise identification of the clinical target volume (CTV) boundary can ensure the operation effect, whereas the CTV boundary is determined by anatomical structures. Therefore, the accurate and automatic segmentation of brain anatomical structures could improve both efficiency and effectiveness for gliomas treatment.

Recently, deep learning approaches have been proven its the superiority on both 2D natural images and 3D medical modalities segmentation task compared to the traditional methods. Especially for the brain tumor segmentation, [5] proposed a U-Net with an additional VAE branch to reconstruct input images and regularize the shared encoder. [4] established a two-stage cascaded U-Net to refine the coarse prediction from the first stage and capture context information in the second stage. As to brain CTV segmentation, [1] introduced DenseNet to predict resection cavity precise contours.

To raise the researcher interest in the study of brain CTV segmentation, the Anatomical Brain Barriers to Cancer Spread challenge (ABCs) aims to encourage challengers to construct an automatic brain structures segmentation methods, where some critical structures (e.g., structures that are served as barriers to the spread of brain cancers or spared from irradiation) are included in two tasks. The dataset provides 45 multi-modal images with ground truth annotations for training and validation, 15 images for the online test, and 15 images for the final test. Each case is given with a CT scan and two diagnostic MRI scan which

© Springer Nature Switzerland AG 2021
N. Shusharina et al. (Eds.): ABCs 2020/L2R 2020/TN-SCUI 2020, LNCS 12587, pp. 27–33, 2021.
https://doi.org/10.1007/978-3-030-71827-5_3

includes contrast-enhanced T1-weighted and T2-weighted FLAIR of the post-operative brain. Besides, all images are co-registered and re-sampled to the size of $164 \times 194 \times 142$ pixels with the isotropic resolution of $1.2 \times 1.2 \times 1.2$ mm. For the task 1, challengers are asked to segment brain structures for the automatic identification of the CTV. As for the task 2, participants are required to segment structures that used in radiotherapy treatment plan optimization. Evaluation metrics are specified to the Dice score and surface Dice score to evaluate the performance of the proposed algorithms.

In this report, we propose a residual block [2] based U-Net which is composed of a deep encoder and a shallow decoder and implemented with nn-UNet framework. To provide multi-scale guidance for boosting framework performance and accelerate training convergence, we employ a deep supervision strategy in our framework. To suppress the false-negative results and retrieve the missing small targets, we utilize the Tversky loss together with the cross-entropy loss as our criterion. Experiments show that the proposed framework outperforms the compared state-of-the-art methods.

2   Method

Based on the U-Net [6] architecture and nn-UNet [3] framework, we propose a variant of this approach based on residual blocks. The details are illustrated as follows.

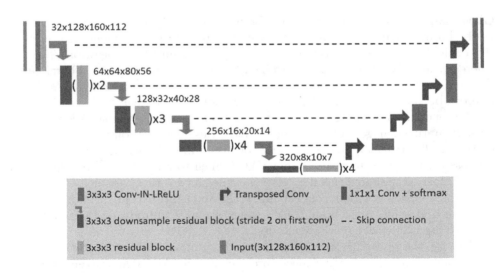

Fig. 1. An overview of the proposed framework.

2.1   Encoder Design

The proposed encoder is composed of residual blocks as shown in Fig. 1, which consist of two $3 \times 3 \times 3$ 3D convolutions layers following with a normalization layer and a activation layer. Then, an identity skip connection operation is employed to link the shallow and the corresponding deep level features. Specifically, we use Instance Normalization (IN) as the normalization layer, since it performs better than BatchNorm when batch size is small (ours is 2), and requires less computation cost than Group Normalization (GN). In addition, We utilize LeakyReLU as the activation layer to retain more information than ReLU. Totally, we have 5 spatial levels with the number of residual blocks in each level of 1, 2, 3, 4, 4. The number of filters, which is implemented by a 3D stridden convolution layer with $3 \times 3 \times 3$ filter and a stride of 2, is initiated with 32. Then, we doubles this number after each downsample operation. Notably, the number of the filter in the last level is set to 320 for computational efficiency. The last downsample convolution avoid the axis z for retaining the slice-wise resolution. We randomly crop the patch with the size of $3 \times 128 \times 160 \times 112$ from 3-modality images as the framework input. After the proposed framework processing, the final output is obtained with a size of $320 \times 8 \times 10 \times 7$.

2.2   Decoder Design

The decoder structure is the opposite operation of the encoder, composed of an upsample block and a standard convolution block, where the upsample block is implemented with a $1 \times 1 \times 1$ 3D convolution layer and a 3D transpose convolution layer. The 3D convolution layer aims to reduce the number of features by a factor of 2, and the 3D transpose convolution layer is used for double the spatial dimension. The standard convolution block consists of a simple $3 \times 3 \times 3$ 3D convolution layer followed with a Instance Normalization and a Leaky ReLU layers. At the end of the decoder, a $1 \times 1 \times 1$ convolution layer with the output channels of target classes conjoins a softmax activation layer to get the final prediction. The shallow decoder could not only boost the speed of training and inference stage but also avoid overfitting.

To accelerate model convergence and provide multi-scale guidance, we also introduce a deepvision strategy, where other three levels of the features will be processed with an extra output block so as to obtain the additional predictions and supervised by the same loss function. The final loss will be weighted differently with respect to the importance of each level output. Here, we set the weights of 8/15, 4/15, 2/15, and 1/15 for the outputs from level 1, 2, 3, and 4 respectively.

2.3   Loss Function

We utilize the DC-CE loss as our criterion, which is composed of origin linear Dice loss and cross-entropy loss:

$$L_{DCCE} = -\frac{\sum_{i=1}^{N}\sum_{c=1}^{C} p_{i,c}y_{i,c} + \epsilon}{\sum_{i=1}^{N}\sum_{c=1}^{C} p_{i,c} + y_{i,c} + \epsilon} - \frac{1}{N}\frac{1}{C}\sum_{i=1}^{N}\sum_{c=1}^{C} y_{i,c}\log(p_{i,c}) \quad (1)$$

where N is the total number of pixels, and C denotes all classes. $p_{i,c}$ and $y_{i,c}$ represent the prediction and annotation of pixel i and class c, respectively. The Tversky loss is also introduced to train our proposed framework, which can be formulated as:

$$L_{Tversky} = \frac{\sum_{i=1}^{N}\sum_{c=1}^{C} p_{i,c}y_{i,c} + \epsilon}{\sum_{i=1}^{N}\sum_{c=1}^{C} p_{i,c}y_{i,c} + \alpha\sum_{i=1}^{N}\sum_{c=1}^{C} p_{i,c}\bar{y}_{i,c} + \beta\sum_{i=1}^{N}\sum_{c=1}^{C} \bar{p}_{i,c}y_{i,c} + \epsilon} \quad (2)$$

where $p_{i,c}\bar{y}_{i,c}$ denotes the false positive (FP) pixels; $bar p_{i,c}y_{i,c}$ denotes the false negative (FN) pixels. The importance of FP and FN is weighted by α and β, which are set to 0.3 and 0.7, respectively.

2.4   Pseudo Training with Model Ensemble

For a further improvement of segmentation accuracy, we gradually employ the model ensemble strategy and pseudo training strategy. At first, we calculate the average of all selected models to get the prediction of the test volumes:

$$P^* = \frac{1}{N}\sum_{i=1}^{N} P_i \quad (3)$$

The P_i denotes the specific prediction of different models, while the P^* denotes the final ensembled prediction. We find it more stable than any single prediction since the error-prone outliers are removed in the ensemble process.

Based on the reliable predictions, we treat them as the pseudo labels for the test volumes. The final loss function for the model training can be described as follow:

$$L_{hybrid}(V, L) = L_{DCCE}(V, L) + L_{Tversky}(V, L) \quad (4)$$

$$L_{final} = L_{hybrid}(V_s, L_s) + L_{hybrid}(V_t, P_t^*) \quad (5)$$

We define the sum of L_{DCCE} and $L_{Tversky}$ as L_{hybrid}, and utilize the ensembled prediction P^* as the pseudo label for test volumes. The pseudo training strategy can bring the model more training pairs, and the model can learn reliable information from pseudo labels. The result in Table 1 proved the effectiveness of the methods in the section.

2.5 Optimization

We utilize the SGD with initial learning rate of $\alpha_0 = 1e - 4$ and momentum of 0.99 to optimize our network. Poly strategy is employed to progressively decrease learning rate according to:

$$\alpha = \alpha_0 * \left(1 - \frac{e}{N_e}\right)^{0.9} \tag{6}$$

where e an iterator of the current epoch; N_e is the total number of the training epochs. We set $N_e = 1000$ with batch size of 2 in the training stage. L2 norm regularization with a weight of $1e - 5$ is used to prevent overfitting.

3 Experiments

3.1 Implementation Details and Data Processing

In this section, we will demonstrate the experimental details of our method. Firstly, we adopt the normalization for three modalities images by subtracting the mean values and dividing the variances. Specifically, to reduce the negative effect from extremum in CT images, the gray intensity in CT modality is clipped into [0.5, 0.995].

Afterwards, we concatenate 3 modalities to obtain a 3-channel volume as the framework input. A series of data augmentation strategies are employed, including rotation, scale, mirror, gamma correction, and brightness additive, to improve the robustness of the framework. As to the inference phase, the test time augmentation strategy (e.g., sliding window and flipping across three axes) is introduced to improve the performance. It is worth noting that adopting horizontal flip along with the x-axis could greatly reduce the accuracy in our experiment. The reason might be that such operation will misleads the framework when learning the paired tissues such as eyes and cochlea. Therefore, we decide to avoid the horizontal flip along with x-axis in Task 2. To evaluate the effectiveness of the proposed framework, we employ the 5-fold cross validations, and choose the proper models to generate final submission. All proposed framework is implemented in PyTorch using an NVIDIA Tesla V100 GPU.

Table 1. Quantitative experiment of the proposed segmentation framework.

Method	Task1 DSC	Task1 SDSC	Task2 DSC	Task2 SDSC
nnU-Net [3]	0.872	0.974	0.777	0.926
ResUnet	0.873	0.973	0.780	0.929
ResUnet+Tversky	0.875	0.973	0.779	0.928
Ensembled pseudo training	**0.876**	**0.974**	**0.782**	**0.929**

3.2 Quantitative and Qualitative Analysis

We evaluate the performance of the proposed frameworks on the online test dataset, with the evaluation metrics as Dice score (DSC) and surface Dice score (SDSC) for Task1 and Task2. The experiment result as shown in Table 1, we choose nnU-Net as the state-of-the-art for comparison, which achieves 87.2% in DSC and 97.4% in SDSC of Task1 and 77.7% in DSC and 92.6% in SDSC of Task2, respectively. In contrast to nn-UNet, our ResUnet achieves the improvement with 0.3% in DSC and 0.3% in SDSC of Task2, and ResUnet with Tversky loss achieves improvement with 0.1% of DSC and 0.1% of SDSC in Task2. To further boost the framework performance, we ensemble the prediction of above methods to get better results.

We visualize the predictions for a qualitative analysis. As shown in Fig. 2 the predictions of our proposed framework are very close to corresponding labels. We also reconstructed the prediction in 3D in Fig. 3, where all target issues are clearly identified.

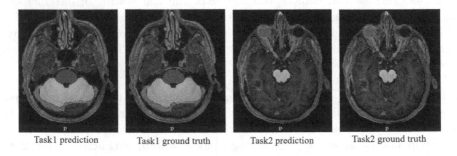

Task1 prediction   Task1 ground truth   Task2 prediction   Task2 ground truth

Fig. 2. Examples of framework's prediction and corresponding ground truth.

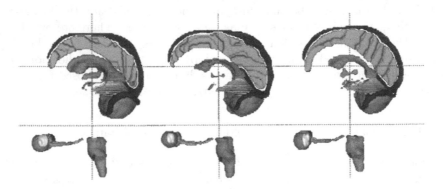

Fig. 3. 3D reconstructions of framework's prediction.

4 Conclusions

In this study, we proposed a effective framework for Anatomical Brain Barriers to Cancer Spread challenge. Specifically, we used residual block based U-shape network as the proposed architecture and the Tversky loss as the criterion, to enforce the feature extraction ability. The ensemble strategy was adopted to refine the prediction and get the better result. Experiments on the online test set varied the efficacy of proposed framework.

References

1. Ermiş, E., et al.: Fully automated brain resection cavity delineation for radiation target volume definition in glioblastoma patients using deep learning. Radiat. Oncol. **15**, 1–10 (2020)
2. He, K., Zhang, X., Ren, S., Sun, J.: Deep residual learning for image recognition. In: Proceedings of the IEEE Conference on Computer Vision and Pattern Recognition, pp. 770–778 (2016)
3. Isensee, F., et al.: nnu-net: self-adapting framework for U-Net-based medical image segmentation. arXiv preprint arXiv:1809.10486 (2018)
4. Jiang, Z., Ding, C., Liu, M., Tao, D.: Two-stage cascaded U-Net: 1st place solution to BraTS challenge 2019 segmentation task. In: Crimi, A., Bakas, S. (eds.) BrainLes 2019. LNCS, vol. 11992, pp. 231–241. Springer, Cham (2020). https://doi.org/10.1007/978-3-030-46640-4_22
5. Myronenko, A.: 3D MRI brain tumor segmentation using autoencoder regularization. In: Crimi, A., Bakas, S., Kuijf, H., Keyvan, F., Reyes, M., van Walsum, T. (eds.) BrainLes 2018. LNCS, vol. 11384, pp. 311–320. Springer, Cham (2019). https://doi.org/10.1007/978-3-030-11726-9_28
6. Ronneberger, O., Fischer, P., Brox, T.: U-Net: convolutional networks for biomedical image segmentation. In: Navab, N., Hornegger, J., Wells, W.M., Frangi, A.F. (eds.) MICCAI 2015. LNCS, vol. 9351, pp. 234–241. Springer, Cham (2015). https://doi.org/10.1007/978-3-319-24574-4_28

An Enhanced Coarse-to-Fine Framework for the Segmentation of Clinical Target Volume

Huai Chen[1], Dahong Qian[2], Weiping Liu[3], Hui Li[4], and Lisheng Wang[1](✉)

[1] Institute of Image Processing and Pattern Recognition, Department of Automation, Shanghai Jiao Tong University, Shanghai 200240, People's Republic of China
`lswang@sjtu.edu.cn`
[2] School of Biomedical Engineering, Shanghai Jiao Tong University, Shanghai 200240, People's Republic of China
[3] Department of Algorithm and Research, Shanghai Aitrox Technology Co., Ltd., Shanghai, China
[4] National Key Laboratory of Science and Technology on Nano/Micro Fabrication, Key Laboratory for Thin Film and Micro Fabrication of the Ministry of Education, Institute of Micro-Nano Science and Technology, Shanghai Jiao Tong University, Shanghai 200240, People's Republic of China

Abstract. In radiation therapy, obtaining accurate boundary of the clinical target volume (CTV) is the vital step to decrease the risk of treatment failures. However, it is a time-consuming and laborious task to obtain the delineation by hand. Therefore, an automatic algorithm is urgently needed to realize accurate segmentation. In this paper, we propose an enhanced coarse-to-fine frameworkto automatically fuse the information of CT, T1 and T2 images to get the target region. This framework includes a coarse-segmentation stage to identify the region of interest (ROI) of targets and a fine-segmentation stage to iteratively refine the segmentation. In the coarse-segmentation stage, the F-loss is proposed to keep the high recall rate of the ROI. In the fine segmentation, the ROI of target will be first cropped according to the ROI obtained by coarse-segmentation and be fed into a 3D-Unet to get the initial results. Then, the prediction and medium features will be set as additional information for the next one network to refine the results. When evaluated on the validation dataset of challenge of Anatomical Brain Barriers to Cancer Spread (ABCs), our method won the 3th place in the public leaderboard.

Keywords: Clinical target volume segmentation · Multi-modality medical images · Coarse-to-fine · Iterative refinement

1 Introduction

In radiation therapy, the delivery of a radiation dose to the treatment target is very conformal, with as low as 1 mm uncertainty [1]. Obtaining accurate boundary of the clinical target volume (CTV) is important for conformal treatments

© Springer Nature Switzerland AG 2021
N. Shusharina et al. (Eds.): ABCs 2020/L2R 2020/TN-SCUI 2020, LNCS 12587, pp. 34–39, 2021.
https://doi.org/10.1007/978-3-030-71827-5_4

to reduce the risk of treatment failures. Recently, radiologists mark the CTV regions slice by slice by hand. It is time-consuming and deeply depends on the experiments of doctors. Therefore, automatic and accurate segmentation of CTV is urgently needed for alleviating the workload of clinicians and improving the efficiency of treatment planning.

The goal of ABCs is to identify the best methods of segmenting brain structures that serve as barriers to the spread of brain cancers and structures to be spared from irradiation, for use in computer assisted target definition for glioma and radiotherapy plan optimization. The segmentation of brain structures is a challenging task as different structures can be visually appreciated more or less favorably on different imaging modalities. Furthermore, as multi-modality images are usually acquired at different time points, they could present subtle anatomical differences even for brain imaging. This presents a unique technological challenge as information from multi-modality imaging is used to define the clinical target volume and the healthy organs for each individual patient's disease [1].

There are totally 15 targets in this challenge, including 5 CTVs and 10 structures. And all of them have various shapes and size. Therefore, the first challenge is to propose an unified framework fitting to all of them. To address this challenge, the coarse-to-fine framework, which firstly adopt coarse segmentation to identify the ROI of target and then adopt fine segmentation to get accurate predictions, is a good choice for the base model. When utilizing the idea of coarse-to-fine, we can not only divide complex segmentation task of 15 targets into simple one-target task but also be successfully process targets with extreme small size.

For the fusion of multi-modal medical images, previous works can be divided into 3 categories according to the fusion style. They are input-level fusion [3,5], medium-level fusion [7,8] and decision-level fusion [4,6]. Input-level fusion based methods directly concatenates the original images, while medium-level fusion based methods fuse features of medium layers and decision-level fusion based methods fuse final outputs of models. In this paper, we utilize the input-level fusion.

Based on the base idea of above analysis, which including utilizing coarse-to-fine framework and input-level fusion, we propose a enhanced coarse-to-fine method. In the coarse-segmentation stage, we propose F-loss to keep high recall rate of ROI to alleviate the missing of target region. And in the fine-segmentation, we propose iterative refinement to iteratively refine the results based on previous predictions and features. Finally, referring to our previous work of [2], we reuse features and results of models in fine-segmentation stage and fuse them to make final predictions. When evaluated on the validation set of challenge of ABCs, our method won the 3th place with mean DSC of 0.883, mean SDSC of 0.980 for task1 and mean DSC of 0.775, mean SDSC of 0.942 for task2.

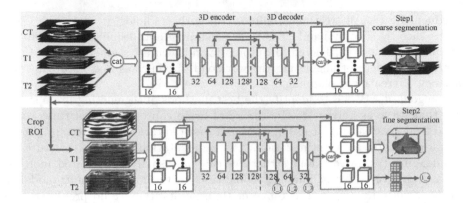

Fig. 1. An illustration of the coarse-to-fine framework.

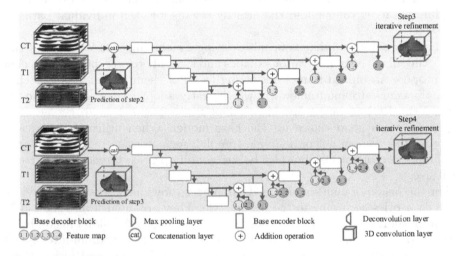

Fig. 2. An illustration of the iterative refinement.

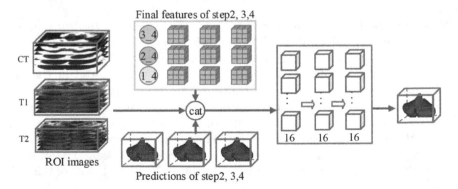

Fig. 3. An illustration of the ensemble refinement.

2   Method

Our framework can be divided into three stages. The first stage is coarse-segmentation, aiming to identify the ROI of targets. The F-loss is proposed to alleviate the missing of important regions by keeping high recall rate. The second stage is fine-segmentation stage, in which, ROI will be firstly cropped and be segmented. In this stage, several 3D-Unets are built, and we propose iterative refinement to repeatedly refine the results by reusing predictions and medium features of previous models in the new one. The third stage is ensemble refinement, where all of the final feature maps and predictions of networks in fine-segmentation stage will be fused and fed into a fusion block to make final decisions. The illustration of these stage are respectively shown in Figs. 1, 2 and 3.

2.1   F-Loss to Keep the High Recall Rate of Coarse Segmentation

In the coarse segmentation stage, keeping high recall rate to alleviate the missing of important regions is the key concern. We define a F-loss loss function to increase recall rates and thereby to alleviate incomplete ROI definitions. The F-loss is inspired by F-score ($F_{score} = (1 + \beta^2) \times \frac{Precision \times Recall}{\beta^2 \times Precision + Recall}$), where setting $\beta > 1$ shows preference for *Recall*. F-loss function is defined as below:

$$
\begin{aligned}
Loss_F(\beta) &= -(1 + \beta^2) \times \frac{\frac{\sum_{i=1}^{N} p_i g_i}{\sum_{i=1}^{N} p_i} \times \frac{\sum_{i=1}^{N} p_i g_i}{\sum_{i=1}^{N} g_i}}{\beta^2 \times \frac{\sum_{i=1}^{N} p_i g_i}{\sum_{i=1}^{N} p_i} + \frac{\sum_{i=1}^{N} p_i g_i}{\sum_{i=1}^{N} g_i}} \\
&= -\frac{(1 + \beta^2) \times (\sum_{i=1}^{N} p_i g_i)}{\beta^2 \times \sum_{i=1}^{N} g_i + \sum_{i=1}^{N} p_i}
\end{aligned}
\tag{1}
$$

We can set $\beta > 1$ ($\beta = 4$ *in this paper*) to keep high recall rate. And it worth noting that it will be the dice loss if $\beta = 1$.

2.2   Iterative Refinement to Iteratively Refine the Results

Iterative refinement is proposed to reuse predictions and features of previous models as additional information to enhance the new model. As shown in Fig. 1 and Fig. 2, the fine segmentation stage of base coarse-to-fine framework, named as step2 in Fig. 1, will firstly obtain the initial prediction and medium features. And then these features and predictions will be as additional information for the next network (named as step3), where predictions will be directly merged into the original images and features will be added into the corresponding features. similarly, previous features of step2 and step3 will be added into step4 and results of step3 will be merged into the input of step4 for step4 iterative refinement.

2.3   Ensemble Refinement to Fuse Multiple Information to Get Finer Results

The final stage is ensemble refinement, in which all previous predictions and features are merged with original images to get finer results. As shown in Fig. 3, all features and predictions from step2, step3 and step4 are merged with original images and fed into a fusion block consisted with 3 convolution layers.

3   Experiments

3.1   DataSet

The challenge of ABCs2020 provides 45 multi-modal images (CT, T1 and T2) with their ground truth annotations as the training data, 15 instances as the validation data and 15 instances as the final test data.

3.2   Training Details

Adam is set as the optimizer for all models and the learning rate is set as 10^{-3}. For the training of coarse-segmentation model, the loss function is F-loss with $\beta = 4$, while for the training of other models, the loss function is F-loss with $\beta = 1$ i.e. dice loss. The total epoch for each model is 50 and we split 5 cases from the training data as the validation data. Therefore, we can set learning rate to half when the validation loss does not decrease in 5 epochs.

4   Results

When evaluated on the validation set, our method obtain mean DSC with 0.883, mean SDSC with 0.980 for task 1, and mean DSC with 0.775, mean SDSC with 0.942 for task 2. The final mean score is 0.888 and is ranked 3th place in the public leaderboard.

References

1. https://abcs.mgh.harvard.edu/index.php
2. Chen, H., Wang, X., Huang, Y., Wu, X., Yu, Y., Wang, L.: Harnessing 2D networks and 3D features for automated pancreas segmentation from volumetric CT images. MICCAI **6**, 339–347 (2019)
3. Havaei, M., et al.: Brain tumor segmentation with deep neural networks. Med. Image Anal. **35**, 18–31 (2017)
4. Kamnitsas, K., et al.: Ensembles of multiple models and architectures for robust brain tumour segmentation. In: Crimi, A., Bakas, S., Kuijf, H., Menze, B., Reyes, M. (eds.) BrainLes 2017. LNCS, vol. 10670, pp. 450–462. Springer, Cham (2018). https://doi.org/10.1007/978-3-319-75238-9_38

5. Kamnitsas, K., Ledig, C., Newcombe, V.F., Simpson, J.P., Kane, A.D., Menon, D.K., Rueckert, D., Glocker, B.: Efficient multi-scale 3D CNN with fully connected CRF for accurate brain lesion segmentation. Med. Image Anal. **36**, 61–78 (2017)
6. Nie, D., Wang, L., Gao, Y., Shen, D.: Fully convolutional networks for multi-modality isointense infant brain image segmentation. In: 2016 IEEE 13th International Symposium on Biomedical Imaging (ISBI), pp. 1342–1345. IEEE (2016)
7. Tseng, K.L., Lin, Y.L., Hsu, W., Huang, C.Y.: Joint sequence learning and cross-modality convolution for 3D biomedical segmentation. In: Proceedings of the IEEE Conference on Computer Vision and Pattern Recognition, pp. 6393–6400 (2017)
8. Valada, A., Mohan, R., Burgard, W.: Self-supervised model adaptation for multi-modal semantic segmentation. arXiv preprint arXiv:1808.03833 (2018)

Automatic Segmentation of Brain Structures for Treatment Planning Optimization and Target Volume Definition

Marco Langhans[1,2,3(✉)], Tobias Fechter[1,2], Dimos Baltas[1,2], Harald Binder[4], and Thomas Bortfeld[1,5]

[1] Division of Medical Physics, Department of Radiation Oncology, Medical Center – University of Freiburg, Faculty of Medicine, University of Freiburg, Freiburg im Breisgau, Germany
marco.langhans@mail.de
[2] German Cancer Consortium (DKTK), Partner Site Freiburg, Freiburg im Breisgau, Germany
[3] Department of Radiation Oncology, Radiologie Vechta, Vechta, Germany
[4] Institute of Medical Biometry and Statistics, Faculty of Medicine and Medical Center, University of Freiburg, Freiburg im Breisgau, Germany
[5] Division of Radiation Biophysics, Department of Radiation Oncology, Massachusetts General Hospital and Harvard Medical School, Boston, MA 02114, USA

Abstract. The MICCAI Challenge 2020 "Anatomical Brain Barriers to Cancer Spread: Segmentation from CT and MR Image" was about segmenting brain structures automatically for further use in the definition of the Clinical Target Volume (CTV) of glioblastoma patients and treatment planning optimization in radiation therapy. This paper describes the methods of the team "FREI". A 3D U-Net style deep learning network was used to achieve human-like segmentation accuracy for most of the structures within seconds.

Keywords: Medical image segmentation · Brain structures · U-Net · Deep learning · MICCAI

1 Introduction

The MICCAI Challenge 2020 "Anatomical Brain Barriers to Cancer Spread: Segmentation from CT and MR Images" consists of two tasks. The first one asks for the automatic segmentation of brain structures of glioblastoma patients for further use in Clinical Target Volume (CTV) definition. The CTV includes a margin of up to 2 cm around the Gross Tumor Volume (GTV), which is the visible part of the tumor. The margin takes the potential presence of non-visible tumor cells into account; it excludes certain structures that are impenetrable

N. Shusharina et al. (Eds.): ABCs 2020/L2R 2020/TN-SCUI 2020, LNCS 12587, pp. 40–48, 2021.
https://doi.org/10.1007/978-3-030-71827-5_5

for tumor cells. Recently [1] described an automated method to define the CTV using an expansion model taking into account brain structures as anatomical barriers. Those barrier structures are falx cerebri, tentorium cerebelli, sagittal and transverse brain sinuses, cerebellum and ventricles. So far these structures have to be segmented manually which is time-consuming and also leads to a high inter-user variability.

Task 2 strives to segment structures (brainstem, eyes, optical nerves, chiasm, lacrimal glands and cochleas) for treatment planning optimization. These structures are so-called organs-at-risk (OAR) that need to be spared from receiving high levels of radiation dose to reduce side effects.

2   Materials and Methods

2.1   Pre-processing

Data Overview. The organizers of the MICCAI 2020 challenge provided a data set from 45 patients for training. Data from 15 additional patients were released as a test set to determine the leaderboard. For the final leaderboard data from another 15 patients were released. For all patients, imaging data consisting of one Computer Tomography (CT) and two Magnetic Resonance Imaging (MRI) scans were provided by the organizers. The two MRI datasets included one T1 and one T2 weighted scan. The labels for Task 1 are cerebellum, falx, sinuses, tentorium and ventricles. They were defined using the T1 weighted MRI.

For Task 2 the labels were defined on the CT scan and consist of the following structures: brainstem, chiasm, cochlea (left), cochlea (right), eye (left), eye (right), lacrimal (left), lacrimal (right), optical nerve (left) and optical nerve (right). The test sets 1 and 2 did not contain any labelmaps.

The available data showed registration errors between CT and MRI and also between the T1 and T2 images. Due to the specifics of how the structures were defined, the CT did not add useful information for Task 1 and the MRI did not add useful information for Task 2. To test if better results could be obtained after better registration of the imaging data sets, an affine registration algorithm, SimpleElastix [2, 3], was applied. The multi-modality approach using CT and MRI (T1 and T2) for both tasks still created bias and resulted in lower accuracy, thus didn't improve the outcome. In consequence, the multi-modality approach has not been pursued. A possible way to solve the registration error problem could be a deformable registration algorithm (e.g. [4]), which will be considered in future work. In this work the authors used a uni-modal approach to solve the task, thus for Task 1 the T1 weighted MRI data and Task 2 the CT was used.

Brain Region. To support the segmentation algorithm to distinguish between regular brain tissue and the respective label masks, an automatic brain segmentation algorithm [5] was used to create the brain mask for each patient individually. The brain mask was added to the existing label maps as a further class. Additionally via thresholding of Hounsfield Units, the remaining patient was classified as a further separate class (body contour). With respect to Task 1 there was no gain in accuracy ascertainable, therefore both additional classes were only added to Task 2.

Additional Pre-processing and Creation of the Training Data. Structures belonging to paired organs (such as the left and right eye) were merged into a single structure (eye) to avoid biases during training. We found that it was easier for the network to accurately detect and segment a structure like the eye, without also specifying it as left or right eye. The body side could be easily determined in a post-processing step using the location of the structure. First attempts without using the merging approach led to a frequent misclassification of the structure's side.

All patients were imported using SimpleITK [6], further handled using the numpy array format and cropped to the non-zero region to keep memory usage low. All image values above the 99th and below the 1st percentile were clipped to avoid biases through single pixels with very high or low values. Afterwards the MR images were normalized by Z-score normalization:

$$z = \frac{x - \mu}{\sigma} \tag{1}$$

with z: normalised intensity, x: intensity, μ: mean value, σ: standard deviation. Necessary metadata was saved into a separate file to maintain data like spacing, coordinates of the non-zero region, direction, origin and original shape. The metadata is needed to transfer the plain arrays back to it's original format. There was no need to re-sample the images because they all had the same pixel size of 1.2 mm.

2.2   Training

Deep Learning Architecture. A 3D U-Net [7,8] based network (see Fig. 1) was utilized within the PyTorch framework. Each level block (yellow) consisted of a Group Normalization [9] + Parametric Rectified Linear Unit (PReLU) [10] + 3D Convolution layer (kernel size = 3, stride = 1, padding = 1) with residual connections [11]. Group Normalization was preferred to Batch Normalization because Batch Normalization turned out to be unstable for small batch sizes (one in this case). The red blocks are down- and the blue blocks upsample layers. The first hidden layer started with 24 channels. During test-time a final Softmax layer (violet) was applied.

Optimization. A regular dice loss function (see Eq. 2) was applied and optimized using the ADAM optimizer. The initial learning rate was 0.001, coefficients used for computing running averages of the gradient and its square were 0.9 and 0.999, respectively. No weight decay was applied.

$$DiceLoss = 1 - \frac{2 \cdot |X \cap Y|}{|X| + |Y|} \tag{2}$$

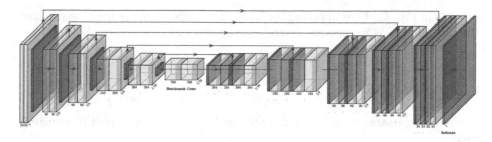

Fig. 1. 3D U-Net. Details see text. Graphic was created using [13] (Color figure online)

with X: Ground Truth and Y: Prediction.

To estimate the performance of the model, a 5-fold cross validation was carried out. 1000 random batches were defined as one epoch. In the first step, the model was trained for 40 epochs. Afterwards the trained model was trained for an additional 40 epochs using the full data set (without validation set) and applied to Test Set 1 and 2 (each n = 15).

Data Loader and Augmentation. Due to the small amount of training data (n = 45), heavy data augmentation was performed using an already implemented data loader [12]. Random cropping, mirroring, scaling, rotations, elastic deformations, gamma transformations, blurring and gaussian noises were applied to the training set. During Training/Validation, patches of $128 \times 128 \times 128$ were randomly cropped from the dataset.

Evaluation. The organizer used two metrices to evaluate the predictions: the commonly known volumetric dice score (see Eq. 2) and the surface dice score. The surface dice score doesn't take the whole volume into account, but the surface volume given a tolerance (2 mm in this challenge). Both metrics ranged between 0 (no overlap) and 1 (perfect overlap).

2.3   Prediction

To predict the final labelmap, a total of 5 patches (128, 128, 128 pixels) were fed into the network. The patches were chosen to sample the whole patient space thus resulting in four quadrants and a center one. Afterwards all labelmaps were merged into one.

2.4   Post-processing

The predictions of both tasks consisted of small false positive dots for structures for a few patients. The removal was done using a "keeping the largest element in an array" method. With respect to Task 2 all false positive dots were removed reliably. Task 1 consists of masks that were partially not connected (e.g. see ventricle in Fig. 3). Thus applying the method resulted in removing true positive volumes as well. As a consequence the method was only applied to all structures of all predictions of Task 2.

3   Results

The following results have been calculated by the organizers as part of the challenge.

Figure 2(a) shows the quantitative results of Task 1. Most of the structures of Task 1 attained mean dice scores over 0.8 (except sinuses with 0.77), which indicates that the segmentation network works well. The cerebellum is perfectly segmented as all patients were close to 1.0, which shows that there is almost no variability. The surface dice scores are, as expected, even better and most of the structures achieved scores above 0.9.

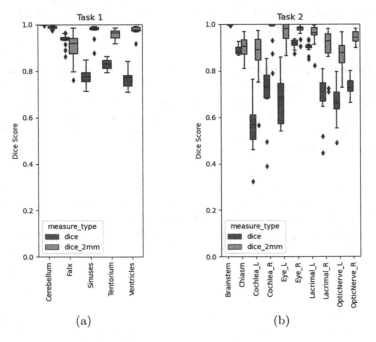

Fig. 2. Volumetric (blue) and surface (orange) dice scores for Task 1 (a) and Task 2 (b). Details see text. (Color figure online)

Figure 2(b) shows results for Task 2. For most structures the results look very good. Structures like the cochleas and lacrimal glands are prone to error due to their relatively small size. Small shifts translate to a large error. It is also noticeable that small structures exhibit a broader spread of both the regular and surface dice score. In general, if the tolerance window is broader (e.g. surface

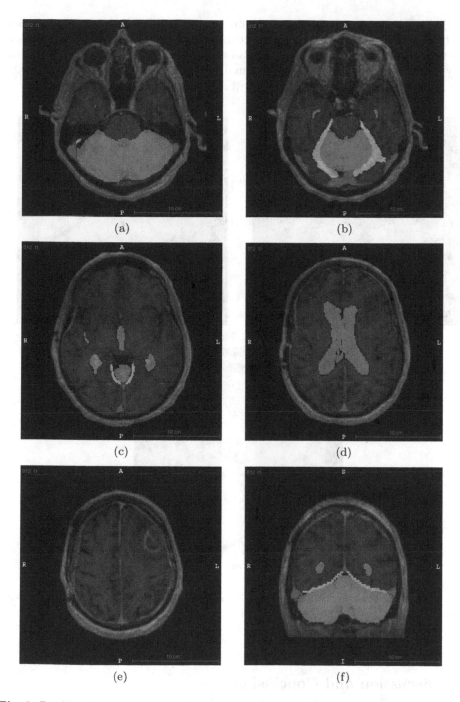

Fig. 3. Predicted segmentations of Task 1 projected on different slices (transversal a-e and coronal f) of the T1 MRI scan. Cerebellum (red), ventricles (cyan), falx cerebri (green), tentorium cerebelli (yellow), sinuses (blue) (Color figure online)

dice) the metric strongly increases. Again, structures like the brainstem, similar to the cerebellum in Task 1, are clearly definable due to their sharp demarcation to other tissue. It is interesting that also the chiasm, which is, especially using only the CT, difficult to segment manually, provides mean dice scores of 0.9. Qualitative examples are shown in Fig. 3 and 4.

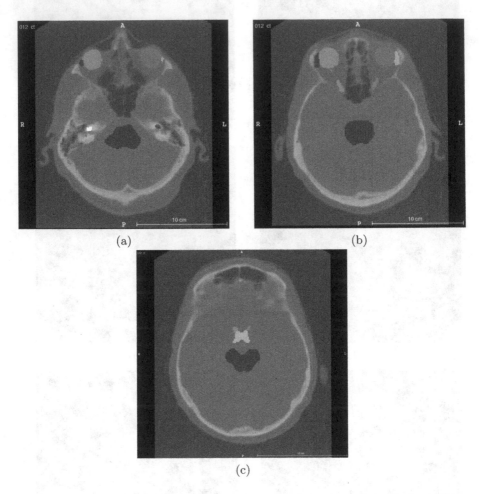

(a)   (b)

(c)

Fig. 4. Predicted segmentations of Task 2 projected on two slices of the CT-scan. Brainstem (red), eyes (magenta, cyan), cochleas (yellow, blue), lacrimal glands (dark blue, white), optical nerves (light and dark brown), chiasm (green) (Color figure online)

4   Discussion and Conclusion

In this work it has been shown that deep neural networks can segment various brain structures in an accurate way. Most of the structures attained volume dice scores of above 0.8, which is approximately the inter-user variability

([14] showed this for comparable brain structures). In other words: the networks achieve human like accuracy.

It is interesting that the dice scores are not symmetrical (with respect to Task 2). One would expect that the left and right hand side of all paired organs are segmented with similar performance, which is not the case. As these structures are also the smallest ones, it leads to the assumption that even slight deviations lead to a large error. Additional investigation will be done in future work.

Further structures like the left eye or the optical nerves, which are clearly appreciable visually, are inferior to structures like the chiasm, which are difficult to identify visually. With respect to the left eye, the surface dice is closer to its right hand side counterpart, which indicates that the geometry itself is captured well, but is segmented slightly too large or small, with respect to the ground truth. Consider a small structure (e.g. a sphere), which has the same shape of the ground truth, but the radius is 2 mm less. Evaluation using the surface dice score would result in a perfect overlap (if a tolerance of 2 mm is given). Although the evaluation using the volumetric dice score would result in a larger error if the volume is small enough.

In this work a uni-modal approach was performed. The structures of Task 2, which are further used for treatment optimization, could be easily derived using the treatment planning CT. No complex registration involving MRI data is needed. In addition, it is more accurate to use the definition of organs at risk on the planning CT, as the calculation of the dose distribution is also done on the CT. Shifts of contours would lead to an under- or overestimation of dose in organs at risk and could even harm the patient. However, the performance may be increased if a deformable registration between CT and MRI is applied. Especially structures that don't show a sharp demarcation in CT, but are appreciable in MRI, may benefit.

Advantages over manual segmentation include a more objective assessment of treatment outcomes, especially with respect to adverse side effects. Because those side effects can dependent on how (correctly or incorrectly) the organs were contoured for treatment planning, a standardized procedure provides more consistency. Also the manual segmentation of a structure set, with various organs at risk, is a time-consuming process. Using the automatic approach the structures are segmented almost instantaneously (within a few seconds).

A limitation of the study is the dependency on the ground truth, as the results strongly depend on how the ground truth is defined. If there are raters providing different styles of how to segment a certain structure it is difficult for the network to learn.

As the split of the merging approach is done using the central axis of the patient it could provide incorrect segmentations for patients showing a deformed brain. Especially with respect to brain tumor patients the presence of deformations is not uncommon.

References

1. Shusharina, N., Soederberg, J., Edmunds, D., Löfman, F., Shih, H., Bortfeld, T.: Automated delineation of the clinical target volume using anatomically constrained 3D expansion of the gross tumor volume. Radiother. Oncol. **146**, 37–43 (2020). https://doi.org/10.1016/j.radonc.2020.01.028
2. Klein, S., Staring, M., Murphy, K., Viergever, M.A., Pluim, J.P.W.: elastix: a toolbox for intensity based medical image registration. IEEE Trans. Med. Imaging **29**(1), 196–205 (2010)
3. Shamonin, D.P., Bron, E.E., Lelieveldt, B.P.F., Smits, M., Klein, S., Staring, M.: Fast parallel image registration on CPU and GPU for diagnostic classification of Alzheimer's disease. Front. Neuroinform. **7**(50), 1–15 (2014)
4. Balakrishnan, G., Zhao, A., Sabuncu, M., Guttag, J., Dalca, A.V.: VoxelMorph: a learning framework for deformable medical image registration. IEEE Trans. Med. Imaging **38**, 1788–1800 (2019)
5. Kleesiek, J., Urban, G., Hubert, A., et al.: Deep MRI brain extraction: a 3D convolutional neural network for skull stripping. Neuroimage **2016**(129), 460–469 (2016)
6. Lowekamp, B.C., Chen, D.T., Ibáñez, L., Blezek, D.: The design of simpleITK. Front. Neuroinform. **7**, 45 (2013)
7. Ronneberger, O., Fischer, P., Brox, T.: U-Net: convolutional networks for biomedical image segmentation. In: Navab, N., Hornegger, J., Wells, W.M., Frangi, A.F. (eds.) MICCAI 2015. LNCS, vol. 9351, pp. 234–241. Springer, Cham (2015). https://doi.org/10.1007/978-3-319-24574-4_28
8. Çiçek, Ö., Abdulkadir, A., Lienkamp, S.S., Brox, T., Ronneberger, O.: 3D U-Net: learning dense volumetric segmentation from sparse annotation. In: Ourselin, S., Joskowicz, L., Sabuncu, M.R., Unal, G., Wells, W. (eds.) MICCAI 2016. LNCS, vol. 9901, pp. 424–432. Springer, Cham (2016). https://doi.org/10.1007/978-3-319-46723-8_49
9. Wu, Y., Kaiming H.: Group normalization. arXiv.orgarxiv.org/abs/1803.08494, 11 June 2018
10. He, K., Zhang, X., Ren, S., Sun, J.: Delving deep into rectifiers: surpassing human-level performance on imagenet classification. In: 2015 IEEE International Conference on Computer Vision (ICCV), Santiago, pp. 1026–1034 (2015). https://doi.org/10.1109/ICCV.2015.123
11. Zhang, Z., Liu, Q., Wang, Y.: Road extraction by deep residual U-Net. IEEE Geosci. Remote Sens. Lett. **15**(5), 749–753 (2018). https://doi.org/10.1109/LGRS.2018.2802944
12. Isensee, F., et al.: batchgenerators - a python framework for data augmentation (2020). https://doi.org/10.5281/zenodo.3632567
13. Iqbal, H.: HarisIqbal88/PlotNeuralNet v1.0.0. (2018)
14. Liedlgruber, M., et al.: Variability issues in automated hippocampal segmentation: a study on out-of-the-box software and multi-rater ground truth. In: 2016 IEEE 29th International Symposium on Computer-Based Medical Systems (CBMS), Dublin, pp. 191–196 (2016)

A Bi-directional, Multi-modality Framework for Segmentation of Brain Structures

Skylar S. Gay$^{(\boxtimes)}$ (iD), Cenji Yu (iD), Dong Joo Rhee (iD), Carlos Sjogreen (iD),
Raymond P. Mumme (iD), Callistus M. Nguyen (iD), Tucker J. Netherton (iD),
Carlos E. Cardenas (iD), and Laurence E. Court (iD)

Department of Radiation Physics, The University of Texas MD Anderson Cancer Center,
Houston, TX 77030, USA
sgay1@mdanderson.org

Abstract. Careful delineation of normal-tissue organs-at-risk is essential for brain tumor radiotherapy. However, this process is time-consuming and subject to variability. In this work, we propose a multi-modality framework that automatically segments eleven structures. Large structures used for defining the clinical target volume (CTV), such as the cerebellum, are directly segmented from T1-weighted and T2-weighted MR images. Smaller structures used in radiotherapy plan optimization are more difficult to segment, thus, a region of interest is first identified and cropped by a classification model, and then these structures are segmented from the new volume. This bi-directional framework allows for rapid model segmentation and good performance on a standardized challenge dataset when evaluated with volumetric and surface metrics.

Keywords: MICCAI 2020 · ABCs · Deep learning · Segmentation

1 Introduction

Successful radiotherapy of brain tumors requires careful identification and delineation of normal organs-at-risk (OARs). Standard clinical practice involves manual delineation, a time-consuming process that is subject to intra- and inter-observer variability [1, 2]. Existing automatic segmentation approaches often fall within four categories: atlas based, machine learning based, deformable based, or region-based. Atlas based approaches involve large amount of registration, a computationally expensive process that is also dependent upon atlas anatomical similarity. Machine learning based techniques suffer from lack of global understanding of shape information, especially for complex structures with inhomogeneous textures. Deformable methods are sensitive to initialization, whereas region-based methods often fail at low contrast boundaries [3].

Recently, deep learning-based methods have demonstrated state-of-the-art performance in a variety of medical image segmentation tasks [4]. Deep neural network architectures such as U-Net [5], SegNet [6], and MeshNet [7] are proposed for segmentation of the brain itself and surrounding structures. However, these approaches often underperform for small structures in the brain whose classes are sufficiently underrepresented

© Springer Nature Switzerland AG 2021
N. Shusharina et al. (Eds.): ABCs 2020/L2R 2020/TN-SCUI 2020, LNCS 12587, pp. 49–57, 2021.
https://doi.org/10.1007/978-3-030-71827-5_6

relatively to the background class. In this work, we therefore apply a two-stage approach to segmenting brain structures. Large structures are directly segmented, while smaller ones are first localized by a classification algorithm, and then segmented. This approach shows good results when used as part of the ABCs [8] challenge, a satellite event of the Medical Image Computing and Computer Assisted Intervention Society (MICCAI) 2020 annual conference.

2   Methods

This approach segments structures from the ABCs dataset by combining predictions from a bi-directional framework. Of the T1-weighted MR, T2-weighted MR, and CT images, only the T1w and T2w MR images are used in this approach to segment large structures. Challenge segmentations are divided into two tasks: Task 1 segments brain structures (falx cerebri, tentorium cerebelli, sagittal and transverse brain sinuses, cerebellum, ventricles) critical for defining clinical target volumes (CTVs) for brain radiotherapy treatment, while Task 2 segments structures (brainstem, optic chiasm, cochleas, eyes, lacrimal glands, and optic nerves) used for optimizing the radiotherapy treatment plan. Large structures, defined as all structures from Task 1 along with the brainstem, and optic chiasm from Task 2, are directly segmented using MR images alone. For small structures (eyes, optic nerves, lacrimal glands, and cochleas), Inception-ResNet-v2 is used to localize each VOI using the CT, followed by segmentation of these structures from the co-registered T1 and T2 VOIs.

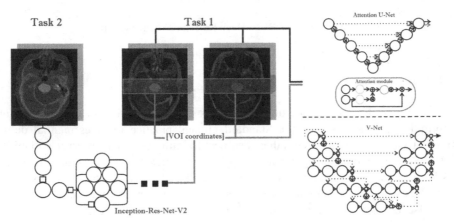

Fig. 1. Bi-directional framework. Task 1 uses T1 and T2 images to directly segment large structures. For Task 2, the CTs are input into an Inception Res-Net-V2 and outputs the range of slices for each small structure (using a single model per structure). These coordinates are used to select VOIs from the T1 and T2 images and are subsequently segmented as in Task 2. Black circles are convolutional blocks, black circles with down/up arrows are down/up sampling, Cyan circles are 1×1 convolutional blocks, green circles are convolutional blocks with ReLU pre-activations and sigmoid post activations, pink squares in the Attention U-Net network are attention gates.

2.1 Data

Data from ABCs [8] were used to train segmentation CNNs. This dataset included glioblastoma and low-grade glioma patients, with all provided patient scans consisted of CT, T1-weighted MR, and T2-weighted MR images of the post-operative brain. All scans were rescaled to have the same spatial dimensions. This dataset provided separate training and testing sets of 45 and 15 patients, respectively. The training dataset was further split 70%–30% training-validation. At the conclusion of the challenge, a previously unseen dataset of 15 additional patients was provided and used for the final scoring stage.

2.2 Training Task 1

After preliminary evaluations of various deep learning models, a modified Attention U-Net [9] with batch normalization and a $3 \times 3 \times 3$ convolutional kernel was selected for training all Task 1 segmentation tasks. DSC was used as the loss function as proposed by Milletari et al. [10]. Adam [11] was used as the optimizer, and the learning rate was set to 0.001 and reduced by 15% when loss did not decrease for 70 epochs. Training was performed on a 16 GB NVIDIA-V100 with batch size 2. Each model was trained for approximately 1,000 iterations using early stopping. Each of the five structures in Task 1 was trained using a dedicated model, and individual structure predictions were combined to create the final predictions.

During training, a body mask was found by thresholding all CT values of –600 or higher. T1w and T2w regions within this mask were normalized by subtracting the mean and then dividing by the standard deviation ("z-score normalization"). Values outside the mask were reset to the minimum value for each scan. Only T1w and T2w images were then used by the model for training.

All data was augmented on-the-fly during training by random combinations of rotations in the x-y plane ($\pm15°$); translations ($\pm10, \pm10, \pm5$) along the x, y, and z axes, respectively; zooms in the x-y plane ($\pm15\%$), and warps. Additionally, before training, all data were flipped left-right to increase the training set size.

2.3 Training Task 2

The structures in Task 2, apart from the brainstem, are significantly smaller than the structures in Task 1, and the approach used for Task 1 was not optimal for the small structures. The relative volume of these structures compared to the input image can be increased by reducing the field of view of a CT scan. To do so, a CNN-based classification was applied to the CT scan to detect the location of the structures, and the images were cropped around the center of mass. The accuracy of the segmentation models improved by training the model with cropped images (i.e. narrow field of view CT scans).

Classification. Inception-ResNet-v2 [12], a CNN-based 2D classification architecture, was trained to classify the existence of an organ-of-interest in each CT slice. The classification model was trained to detect the eyes and the cochleae in each CT slices. The center of mass of each organ in cranial-caudal direction was determined by taking the average of the results from the classification model. Then, 20 slices above and below from the central slice were clipped from an original CT scan to cover the entire organ with some margin. In the segmentation phase, the total of the 40 slices were transferred to the segmentation model. The cropped slices for the eyes were also used to segment the optic nerves, the optic chiasm, and the lacrimal glands, as these structures were all located within the slices cropped for the eyes.

Segmentation. After preliminary evaluations of various deep learning models, two models were selected for Task 2 segmentation based upon individual structure segmentation performance. Each of the six structures in Task 2 was trained using a dedicated model, and individual structure predictions were combined to create the final predictions. Training was performed on a 32 GB NVIDIA-V100 for all models.

The brainstem, eyes, lacrimals, and optic nerves were segmented with a modified Attention U-Net [9] with batch normalization and a $3 \times 3 \times 3$ convolutional kernel. DSC was used as the loss function as proposed by Milletari et al. [10]. Adam [11] was used as the optimizer, and the learning rate was set to 0.001 and reduced by 15% when loss did not decrease for 70 epochs. Batch size was set to 5. Each model was trained for approximately 1,000 iterations using early stopping.

For the remaining Task 2 structures, the optic chiasm and the cochleas, a modified V-Net [10] with batch normalization and a $3 \times 3 \times 3$ convolutional kernel was selected for segmentation. Batch size was set to 2, and all other training hyperparameters followed those described above. The choice of this architecture for these structures, instead of the Attention U-Net architecture used for all other Task 2 segmentations, was based upon visual inspection of preliminary results which revealed better performance for these structures.

During training, a body mask was found by thresholding all CT values of −600 or higher. T1w and T2w regions within this mask were normalized by subtracting the mean and then dividing by the standard deviation ("z-score normalization"). Values outside the mask were reset to the minimum value for each scan. Only T1w and T2w images were then used by the model for training.

All data was augmented on-the-fly during training by random combinations of rotations in the x-y plane ($\pm15°$); translations (±10, ±100, ±5) along the x, y, and z axes, respectively; zooms in the x-y plane ($\pm15\%$), and warps. Additionally, before training all data was flipped left-right to increase the training set size.

2.4 Evaluation Metrics

All submissions to the challenge were evaluated with two metrics: the Dice similarity coefficient (DSC) as well as the surface Dice similarity coefficient (SDSC) with a tolerance of 2 mm [13]. DSC and SDSC scores for individual structures were computed separately, and then averaged for the overall Task 1 and Task 2 scores. Finally, the overall challenge ranking was based upon the unweighted average of the Task 1 and Task 2 scores [8].

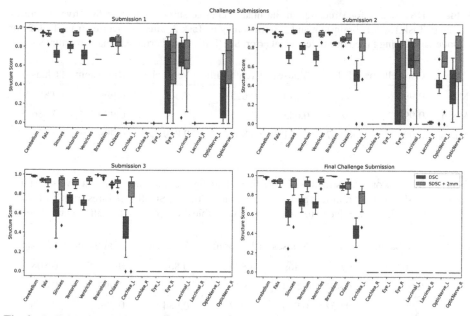

Fig. 2. Individual structure scores for all submissions to the ABCs challenge. DSC is volumetric dice similarity coefficient, and SDSC + 22 mm is surface dice similarity coefficient with 2 mm tolerance.

3 Results

3.1 Challenge Results

In the final challenge submission, relatively good results were achieved for large structure segmentations, with mean DSC above 0.9 for three of the seven large structures, and above 0.7 for three of the remaining. Similarly, mean SDSC with a 2mm tolerance was above 0.9 for six of these. For both metrics, sagittal and transverse brain sinus segmentation scored the lowest, at 0.64 and 0.89 for mean DSC and mean SDSC, respectively.

Conversely, small structure scores were low in the final submission set. This was due to an error in the inference and data conversion scripts, which assigned the left structure class to right structures, and vice-versa. Post-challenge, these errors were corrected, with significant improvement to model performance.

4 Post-challenge Processing

After results were reported at the conclusion of the challenge, poor performance was observed in the segmentation scores for the cochleas, eyes, lacrimal glands, and optic nerves. This was identified as an error in the inference pipeline, which led to organs being incorrectly predicted on the wrong side of the body (i.e. the left eye segment was predicted on the right side of the body, and vice-versa). Performance improved

Table 1. DSC scores for structures in the final challenge submission. The first five structures belong to Task 1, while brainstem and chiasm belong to Task 2. Small structures from Task 2 are excluded as scoring failed due to technical error. See Table 3 for corrected small structure scores.

	Cerebellum	Falx	Sinuses	Tentorium	Ventricles	Brainstem	Chiasm
Min	0.996	0.919	0.244	0.623	0.602	0.999	0.85
Mean	0.997	0.941	0.643	0.726	0.704	0.999	0.886
Max	0.998	0.957	0.751	0.805	0.819	0.999	0.921

Table 2. SDSC scores for structures in the final challenge submission. The first five structures belong to Task 1, while brainstem and chiasm belong to Task 2. Small structures from Task 2 are excluded as scoring failed due to technical error. See Table 4 for corrected small structure scores.

	Cerebellum	Falx	Sinuses	Tentorium	Ventricles	Brainstem	Chiasm
Min	0.969	0.879	0.469	0.834	0.865	0.988	0.808
Mean	0.982	0.934	0.894	0.92	0.942	0.993	0.9
Max	0.99	0.967	0.979	0.987	0.984	0.996	0.966

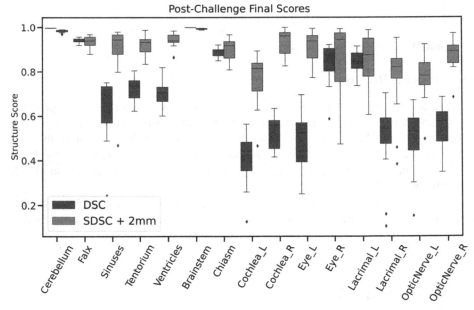

Fig. 3. Individual structure scores after correcting post-processing. DSC is volumetric dice similarity coefficient, and SDSC + 2 mm is surface dice similarity coefficient with 2 mm tolerance.

upon reassigning the structures to the correct orientation. Corrected scores for small structures are listed in Tables 3 and 4.

Table 3. DSC scores for structures corrected post-challenge. These structures, all belonging to Task 2, were unable to be scored during the challenge due to technical error.

	Cochlea L	Cochlea R	Eye L	Eye R	Lacrimal L	Lacrimal R	Optic Nerve L	Optic Nerve R
Min	0.127	0.417	0.251	0.587	0.736	0.107	0.154	0.349
Mean	0.408	0.529	0.488	0.84	0.839	0.504	0.491	0.547
Max	0.564	0.634	0.696	0.922	0.914	0.703	0.671	0.686

Table 4. SDSC scores for structures corrected post-challenge. These structures, all belonging to Task 2, were unable to be scored during the challenge due to technical error.

	Cochlea L	Cochlea R	Eye L	Eye R	Lacrimal L	Lacrimal R	Optic Nerve L	Optic Nerve R
Min	0.466	0.826	0.773	0.474	0.606	0.384	0.497	0.684
Mean	0.77	0.932	0.911	0.844	0.84	0.773	0.778	0.872
Max	0.893	0.999	0.995	0.993	0.987	0.952	0.923	0.972

5 Discussion

Differences in modality make direct comparison between the approach of this study and others' difficult. In particular, published results often focus on MR [14, 15] or CT [13, 16, 17] exclusively instead of the multi-modality approach described here. However, our results generally compare favorably to those published in the literature. For example, this work finds mean brainstem DSC and SDSC as 0.999 and 0.993, respectively, which consistently ranks higher than other reported values [13–17]. Similar improvements for DSC are observed for cerebellum, optic chiasm, and left lacrimals [13–15]. However, SDSC scores for cochleas, lacrimals, and optic nerves were often lower than reported values [13].

There are a few limitations associated with this work. The DSC and SDSC reported for small structures are often, though not always, lower than those generated by CT-only approaches. This may be due to the poorer visibility of these structures in MR imagery, which were the only inputs into the segmentation models of this work. Future research will incorporate CT images into the segmentation models for additional context. In addition, the relatively small test set (n = 15) may be considered a limitation; however, the similarity of our results to those with significantly larger test sets indicates that the size of this test set did not significantly impact the work.

6 Conclusion

In this work, we develop a multi-modality framework for rapid autosegmentation of brain structures as part of an international segmentation challenge. The results of this study showed good agreement with or improvement upon similar values reported in the literature. Future work will include incorporating CT images directly into segmentation models for challenging small structures. This approach could contribute to the Radiation Planning Assistant [18], a fully automated treatment planning tool aimed at improving access to high quality radiation therapy across the globe.

Acknowledgements. The authors acknowledge the support of the High Performance Computing facility at the University of Texas MD Anderson Cancer Center and the Texas Advanced Computing Center (TACC) at The University of Texas at Austin for providing computational resources that have contributed to the research results reported in this paper.

References

1. Jeanneret-Sozzi, W., et al.: The reasons for discrepancies in target volume delineation: a SASRO study on head-and-neck and prostate cancers. Strahlenther. Onkol. **182**(8), 450–457 (2006). https://doi.org/10.1007/s00066-006-1463-6
2. Brouwer, C.L., et al.: 3D variation in delineation of head and neck organs at risk. Radiat. Oncol. **7**, 32 (2012). https://doi.org/10.1186/1748-717X-7-32
3. González-Villà, S., Oliver, A., Valverde, S., Wang, L., Zwiggelaar, R., Lladó, X.: A review on brain structures segmentation in magnetic resonance imaging (2016). https://doi.org/10.1016/j.artmed.2016.09.001
4. Cardenas, C.E., Yang, J., Anderson, B.M., Court, L.E., Brock, K.B.: Advances in Auto-Segmentation (2019). https://doi.org/10.1016/j.semradonc.2019.02.001
5. Ronneberger, O., Fischer, P., Brox, T.: U-Net: convolutional networks for biomedical image segmentation. In: Navab, N., Hornegger, J., Wells, W.M., Frangi, A.F. (eds.) MICCAI 2015. LNCS, vol. 9351, pp. 234–241. Springer, Cham (2015). https://doi.org/10.1007/978-3-319-24574-4_28
6. De Brébisson, A., Montana, G.: Deep neural networks for anatomical brain segmentation. In: IEEE Computer Society Conference on Computer Vision and Pattern Recognition Workshops, pp. 20–28 (2015). https://doi.org/10.1109/CVPRW.2015.7301312.
7. McClure, P., et al.: Knowing What You Know in Brain Segmentation Using Bayesian Deep Neural Networks (2019). https://doi.org/10.3389/fninf.2019.00067. https://www.frontiersin.org/articles/10.3389/fninf.2019.00067/full
8. Shusharina, N., Bortfeld, T., Cardenas, C.E.: MICCAI 2020 ABCs Challenge. https://abcs.mgh.harvard.edu/index.php.
9. Oktay, O., et al.: Attention U-Net: Learning where to look for the pancreas. arXiv. (2018)
10. Milletari, F., Navab, N., Ahmadi, S.A.: V-Net: Fully convolutional neural networks for volumetric medical image segmentation. In: Proceedings - 2016 4th International Conference on 3D Vision, 3DV 2016, pp. 565–571 (2016). https://doi.org/10.1109/3DV.2016.79
11. Kingma, D.P., Ba, J.: Adam: A Method for Stochastic Optimization (2014)
12. Szegedy, C., Ioffe, S., Vanhoucke, V., Alemi, A.A.: Inception-v4, inception-ResNet and the impact of residual connections on learning. In: 31st AAAI Conference on Artificial Intelligence, AAAI 2017, pp.. 4278–4284 (2017)
13. Nikolov, S., et al.: Deep learning to achieve clinically applicable segmentation of head and neck anatomy for radiotherapy. arXiv (2018)
14. Moeskops, P., et al.: Evaluation of a deep learning approach for the segmentation of brain tissues and white matter hyperintensities of presumed vascular origin in MRI. NeuroImage Clin. **17**, 251–262 (2018). https://doi.org/10.1016/j.nicl.2017.10.007

15. Luna, M., Park, S.H.: 3D Patchwise U-Net with transition layers for MR brain segmentation. In: Crimi, A., Bakas, S., Kuijf, H., Keyvan, F., Reyes, M., van Walsum, T. (eds.) BrainLes 2018. LNCS, vol. 11383, pp. 394–403. Springer, Cham (2019). https://doi.org/10.1007/978-3-030-11723-8_40

16. Wang, Y., Zhao, L., Wang, M., Song, Z.: Organ at risk segmentation in head and neck CT images using a two-stage segmentation framework based on 3D U-net. IEEE Access. **7**, 144591–144602 (2019). https://doi.org/10.1109/ACCESS.2019.2944958

17. Rhee, D.J., et al.: Automatic detection of contouring errors using convolutional neural networks. Med. Phys. **46**, 5086–5097 (2019). https://doi.org/10.1002/mp.13814

18. Court, L.E., et al.: Radiation planning assistant - a streamlined, fully automated radiotherapy treatment planning system. J. Vis. Exp. **2018**, e57411 (2018). https://doi.org/10.3791/57411

L2R – Learn2Reg: Multitask and Multimodal 3D Medical Image Registration

Large Deformation Image Registration with Anatomy-Aware Laplacian Pyramid Networks

Tony C. W. Mok$^{(\boxtimes)}$ and Albert C. S. Chung

Lo Kwee-Seong Medical Image Analysis Laboratory,
Department of Computer Science and Engineering,
The Hong Kong University of Science and Technology,
Clear Water Bay, Hong Kong
cwmokab@connect.ust.hk

Abstract. Deep learning-based methods have recently demonstrated remarkable results in deformable image registration for a wide range of medical image analysis tasks. However, most of the deep learning-based approaches are often limited to small deformation settings. In this paper, we describe a deformable image registration approach for the Learn2Reg 2020 challenge based on the Laplacian pyramid image registration networks. Our approach won 1st place in the Learn2Reg 2020 challenge.

Keywords: Image registration · Diffeomorphic registration · Deep Laplacian pyramid networks · Learn2Reg

1 Introduction

Medical image registration is important in a variety of medical image analysis and has been a topic of active research for decades. Recently, several unsupervised deep learning-based approaches [2,3,11,16,19] have been proposed for deformable image registration and achieved remarkable performance in terms of registration accuracy, computation speed and robustness. To evaluate state-of-the-art methods for image registration, the Learn2Reg 2020 challenge [1] consists of four clinically relevant sub-tasks, including brain intra-operative ultrasound (iUS) to MRI registration [17], exhale-to-inhale lung CT registration [9], abdominal CT registration [18], and Hippocampus registration [15]. The clinical sub-tasks in the Learn2Reg challenge impose three critical challenges: estimating large deformations, learning from small datasets, and dealing with multi-modal scans, which are challenging for deep learning-based approaches.

In this paper, we present an image registration method based on a specific deep convolutional neural network (CNN) architecture for large deformation registration, which won the Learn2Reg 2020 challenge. We adopt the Laplacian pyramid network with anatomical label supervision to overcome the large inter- or intra-variations of the anatomical structures in the input scans, and data augmentation to alleviate the overfitting issue.

N. Shusharina et al. (Eds.): ABCs 2020/L2R 2020/TN-SCUI 2020, LNCS 12587, pp. 61–67, 2021.
https://doi.org/10.1007/978-3-030-71827-5_7

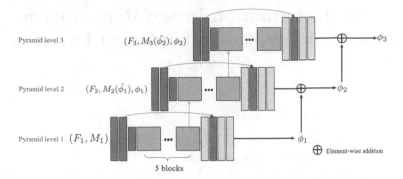

Fig. 1. Overview of the proposed 3-level deep Laplacian pyramid image registration networks. The feature maps from feature encoder, a set of 5 residual blocks, and feature decoder are colored with blue, gray and green, respectively. We highlight that all registrations are done in 3D throughout this paper. For clarity and simplicity, we depict the 2D formulation of our method in the figure. (Color figure online)

2   Methods

Existing deep learning-based approaches [2,3,14,16] often rely on the affine pre-registration and are limited to small deformation settings. Consequently, the registration accuracy degrades whenever there exists large inter- or intra-variation of the anatomical structures in the input scans. Motivated by the successes of deep Laplacian pyramid networks in a variety of computer vision tasks [4,10], we proposed the Laplacian pyramid image registration network (LapIRN) [12] that aims to address the medical image registration problem in large deformation settings.

2.1   Laplacian Pyramid Image Registration Networks

Let F and M denote a fixed 3D scan and a moving 3D scan, respectively. We formulate the image registration problem as a learning problem based on deep CNN. The goal of our method is to estimate the optimal displacement fields ϕ^* such that the dissimilarity between the warped moving scan $M(\phi^*)$ and F are minimized, subject to the smoothness regularization on ϕ^*.

We implement our image registration approach by using a 3-level Laplacian pyramid framework to naturally mimic the conventional multi-resolution strategy. Figure 1 depicts the network architecture of our method. We first create the input image pyramid by downsampling the input images with trilinear interpolation to obtain $F_i \in \{F_1, F_2, F_3\}$ (and $M_i \in \{M_1, M_2, M_3\}$), where F_i denotes the downsampled F with a scale factor $0.5^{(L-i)}$ such that F_1 has the coarsest spatial resolution in F_i and $F_3 = F$. For each level $i \in 1, 2, 3$, we utilize the identical CNN-based registration network (CRN) presented in [12] to capture hierarchical features of the input image pair in pyramid level i (i.e., the concatenation of F_i and M_i) necessary to estimate the correspondence deformation

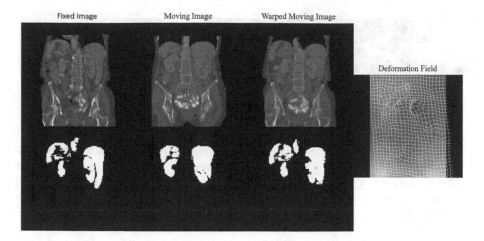

Fig. 2. Example coronal CT slices of inter-patient abdominal CT registration using LapIRN. The corresponding segmentation label maps of the anatomical structures are represented in the second row. Note that we visualize all distinct anatomical labels with the same color for simplicity. (Color figure online)

field ϕ_i at level i. Similar to the conventional multi-resolution optimization approach, the registration starts with the input image pair that has the coarsest spatial resolution (i.e., F_1 and M_1). The resulting deformation field ϕ_1 and the high-level embedding from the CRN in pyramid level i is then passed to the next level. The skip connection between the CRNs at different pyramid levels greatly increases the receptive field as well as the non-linearity of the network to learn complex non-linear correspondence at the finer levels. For pyramid level $i > 1$, the CRN focuses on the misalignment of F_i and $M_i(\hat{\phi}_{i-1})$,and takes the concatenation of F_i, M_i and the upsampled ϕ_{i-1} as input. The output deformation ϕ_i is formed by the addition of the output of CRN in the pyramid level i and added with upsampled deformation field from the previous level $\hat{\phi}_{i-1}$. We repeat this registration process until the finest level completed.

2.2 Anatomical Label Supervision

To overcome the registration difficulties caused by the large inter- or intra-variations of the anatomical structures in the input scans, we introduce the anatomical label supervision along with the similarity pyramid to the loss function of LapIRN. The complete loss is therefore:

$$\mathcal{L}_p = \mathcal{S}^p + \mathcal{L}_{dice} + \frac{\lambda}{2^{(L-p)}} ||\nabla \phi_p||_2^2, \tag{1}$$

where $p \in \{1, 2, 3\}$ denotes the current pyramid level, $\mathcal{S}^p$ represents the similarity pyramid with p levels, $\mathcal{L}_{dice}$ is the Dice score loss of the anatomical segmentation labels, the last term is the smoothness regularization on the displacement fields

Fixed Image   Fixed Label   Moving Image   Moving Label   Warped Moving Image   Warped Moving Label   Deformation Field

Fig. 3. Example sagittal MR slices of Hippocampus registration using LapIRN.

ϕ_p, and λ is a regularization parameter. The similarity pyramid is a weighted sum of the similarity between the fixed image and warped moving image in different spatial resolutions, which is less sensitive to the image noise and helps to avoid local minimal solutions. Mathematically, The proposed similarity pyramid is formulated as:

$$\mathcal{S}^K(F, M) = \sum_{i \in [1..K]} -\frac{1}{2^{(K-i)}} NCC_w(F_i, M_i), \tag{2}$$

where $\mathcal{S}^K(\cdot, \cdot)$ denotes the similarity pyramid with K levels, NCC_w represents the local normalized cross-correlation with windows size w^3, and (F_i, M_i) denotes the images in the image pyramid. Although the similarity pyramid is robust to the noise, it heavily relies on the intensity-based similarity metric, which implies that this metric cannot differentiate distinct anatomical structures with similar intensity values presented in the input scans. As such, we decide to utilize the anatomical segmentation label in the dataset and introduce an anatomy-aware loss term $\mathcal{L}_{dice}$. Given the anatomical segmentation labels of the fixed image F_i^{seg} and the moving image M_i^{seg} in i-th pyramid level. The proposed anatomy-aware loss term is formulated as:

$$\mathcal{L}_{dice}(F_i^{seg}, M_i^{seg}(\phi_i)) = \frac{2|F_i^{seg} \cap M_i^{seg}(\phi_i)|}{|F_i^{seg}| + |M_i^{seg}(\phi_i)|}. \tag{3}$$

2.3 Data Preprocessing and Augmentation

We adopt a simple preprocessing pipeline for the image scans in all sub-tasks. We downsample all the image scans with a factor of 2 and normalized the intensity values to $[0, 1]$. For task 3, we also apply the windowing with lower and upper bound set to -500 and 800 respectively on the abdominal CT scans. During the training phase of our method, we apply affine augmentation with a probability of 0.1 to our training data in order to alleviate the overfitting issue. Specifically, the affine matrix used to augment the training data is defined as: $A = R_x R_y R_z T_{xyz}$, where R_x, R_y and R_z denotes the rotation matrices over x, y and z axis respectively. T_{xyz} represents the composition of a translation and scaling matrices. We randomly sample the degree of rotation, translation and scaling parameters from $[-10, 10]$, $[-0.05, 0.05]$ and $[-0, 08, 0.08]$.

3 Results

We implemented our method in Pytorch [13] and trained it on an NVIDIA RTX2080TI GPU using the Learn2Reg 2020 training dataset. As our method is not capable of multi-modal registration, we use a conventional affine registration approach (6 DoF) with the Normalized gradient field as the similarity metric and the Limited-memory the L-BFGS as the optimizer for task 1. Table 1 shows the results of our model and the other top-performance teams on the Learn2Reg 2020 testing dataset. Our method achieves the best overall ranking among ten international groups. In particular, our method shows the state-of-the-art results in abdominal CT registration (Task 3) and Hippocampus MR registration (Task 4) in terms of the dice similarity coefficient, Hausdorff distance of segmentation, smoothness of the solutions and runtime. Qualitative results of abdominal CT registration and Hippocampus MR registration on cases with large inter-variation are depicted in Figs. 2 and 3. The result in Fig. 2 shows that not only does our method achieve promising registration accuracy on the labeled anatomical structures under the large deformation setting, our method is also capable to register the unlabeled regions, including spinal vertebra and pelvis.

Table 1. Learn2Reg 2020 testing dataset results. Average target registration error of landmarks (TRE), Dice score (DSC), Standard deviation of log Jacobian determinant of the deformation field (SDlogJ) and runtime in second(s) (Time) for each task.

Method	Task1			Task2			Task3			Task4		
	TRE	SDlogJ	Time	TRE	SDlogJ	Time	DSC	SDlogJ	Time	DSC	SDlogJ	Time
Initial	6.38	-	-	10.24	-	-	0.23	-	-	0.55	-	-
Ours	5.67	0.00	31.21	3.24	0.06	1.33	0.67	0.12	1.83	0.88	0.05	1.00
PDD-Net [5,6]	3.08	0.00	4.61	2.46	0.04	2.49	0.46	0.43	4.05	0.78	0.07	0.31
deeds [7,8]	3.93	0.00	9.12	2.26	0.07	41.32	0.51	0.11	41.65	0.76	0.11	3.14

4 Conclusion

In this paper, we have described the method that we used in the Learn2Reg 2020 challenge. By integrating the conventional multi-resolution optimization strategy with deep neural network, our method inherits the runtime advantage of the deep neural network and achieves the state-of-the-art results in most of the clinically relevant sub-tasks in the Learn2Reg 2020 challenge.

References

1. Learn2reg: 2020 miccai registration challenge. https://learn2reg.grand-challenge.org/. Accessed 10 Nov 2020

2. Balakrishnan, G., Zhao, A., Sabuncu, M.R., Guttag, J., Dalca, A.V.: An unsupervised learning model for deformable medical image registration. In: Proceedings of the IEEE Conference on Computer Vision and Pattern Recognition, pp. 9252–9260 (2018)
3. Dalca, A.V., Balakrishnan, G., Guttag, J., Sabuncu, M.R.: Unsupervised learning for fast probabilistic diffeomorphic registration. In: Frangi, A.F., Schnabel, J.A., Davatzikos, C., Alberola-López, C., Fichtinger, G. (eds.) MICCAI 2018, Part I. LNCS, vol. 11070, pp. 729–738. Springer, Cham (2018). https://doi.org/10.1007/978-3-030-00928-1_82
4. Ghiasi, G., Fowlkes, C.C.: Laplacian pyramid reconstruction and refinement for semantic segmentation. In: Leibe, B., Matas, J., Sebe, N., Welling, M. (eds.) ECCV 2016, Part III. LNCS, vol. 9907, pp. 519–534. Springer, Cham (2016). https://doi.org/10.1007/978-3-319-46487-9_32
5. Heinrich, M.P., Hansen, L.: Highly accurate and memory efficient unsupervised learning-based discrete CT registration using 2.5D displacement search. In: Martel, A.L., et al. (eds.) MICCAI 2020, Part III. LNCS, vol. 12263, pp. 190–200. Springer, Cham (2020). https://doi.org/10.1007/978-3-030-59716-0_19
6. Heinrich, M.P., et al.: Mind: modality independent neighbourhood descriptor for multi-modal deformable registration. Med. Image Anal. **16**(7), 1423–1435 (2012)
7. Heinrich, M.P., Jenkinson, M., Brady, M., Schnabel, J.A.: MRF-based deformable registration and ventilation estimation of lung CT. IEEE Trans. Med. Imaging **32**(7), 1239–1248 (2013)
8. Heinrich, M.P., Maier, O., Handels, H.: Multi-modal multi-atlas segmentation using discrete optimisation and self-similarities. VISCERAL Challenge@ ISBI 1390, 27 (2015)
9. Hering, A., Murphy, K., Ginneken, B.V.: Lean2Regchallenge: CT lung registration-training data [data set]. Zenodo (2020)
10. Lai, W.S., Huang, J.B., Ahuja, N., Yang, M.H.: Fast and accurate image super-resolution with deep Laplacian pyramid networks. IEEE Trans. Pattern Anal. Mach. Intell. **41**(11), 2599–2613 (2018)
11. Mok, T.C., Chung, A.: Fast symmetric diffeomorphic image registration with convolutional neural networks. In: Proceedings of the IEEE/CVF Conference on Computer Vision and Pattern Recognition, pp. 4644–4653 (2020)
12. Mok, T.C.W., Chung, A.C.S.: Large deformation diffeomorphic image registration with Laplacian pyramid networks. In: Martel, A.L., et al. (eds.) MICCAI 2020, Part III. LNCS, vol. 12263, pp. 211–221. Springer, Cham (2020). https://doi.org/10.1007/978-3-030-59716-0_21
13. Paszke, A., Gross, S., Chintala, S., et al.: Automatic differentiation in pytorch. In: NIPS-W (2017)
14. Rohé, M.-M., Datar, M., Heimann, T., Sermesant, M., Pennec, X.: SVF-Net: learning deformable image registration using shape matching. In: Descoteaux, M., Maier-Hein, L., Franz, A., Jannin, P., Collins, D.L., Duchesne, S. (eds.) MICCAI 2017. LNCS, vol. 10433, pp. 266–274. Springer, Cham (2017). https://doi.org/10.1007/978-3-319-66182-7_31
15. Simpson, A.L., et al.: A large annotated medical image dataset for the development and evaluation of segmentation algorithms. arXiv preprint arXiv:1902.09063 (2019)
16. de Vos, B.D., Berendsen, F.F., Viergever, M.A., Staring, M., Išgum, I.: End-to-end unsupervised deformable image registration with a convolutional neural network. In: Cardoso, M., et al. (eds.) DLMIA/ML-CDS -2017. LNCS, vol. 10553, pp. 204–212. Springer, Cham (2017). https://doi.org/10.1007/978-3-319-67558-9_24

17. Xiao, Y., Fortin, M., Unsgård, G., Rivaz, H., Reinertsen, I.: REtrospective evaluation of cerebral tumors (resect): a clinical database of pre-operative MRI and intra-operative ultrasound in low-grade glioma surgeries. Med. Phys. **44**(7), 3875–3882 (2017)
18. Xu, Z., et al.: Evaluation of six registration methods for the human abdomen on clinically acquired CT. IEEE Trans. Biomed. Eng. **63**(8), 1563–1572 (2016)
19. Zhao, S., Dong, Y., Chang, E.I., Xu, Y., et al.: Recursive cascaded networks for unsupervised medical image registration. In: Proceedings of the IEEE International Conference on Computer Vision, pp. 10600–10610 (2019)

Discrete Unsupervised 3D Registration Methods for the Learn2Reg Challenge

Lasse Hansen$^{(\boxtimes)}$ ⓘ and Mattias P. Heinrich ⓘ

Institute of Medical Informatics, Universität zu Lübeck, Lübeck, Germany
{hansen,heinrich}@imi.uni-luebeck.de

Abstract. The Learn2Reg challenge poses four very different tasks with varying difficulty for image registration algorithms. In this short paper, we describe our choices for two state-of-the-art discrete 3D registration methods that enable fast and accurate estimation of large deformations without expert supervision during training. Both approaches primarily focus on the use of contrast-invariant features with dense displacement evaluation and were ranked among the top three of all challenge contestants, yielding two first places and three second places for the four sub-tasks.

Keywords: Discrete optimisation · Graphical models · Constrast-independent features

1 Motivation and Background

Deformable image registration, in particular deep learning based approaches, have often struggled with capturing large deformations of local anatomy. Conventional registration methods either rely on multiple warps and coarse-to-fine resolution schemes, which are hard to mimic within learning based framework due to memory constraints, or discrete optimisation to avoid local minima. Here, we analyse our recently proposed probabilistic dense displacement net (**PDD-net**) [4,5] in comparison to the state-of-the-art 3D discrete registration framework **deeds** [7] using modality-invariant self-similarity descriptors [6,9]. Our results demonstrate that despite using no labels for supervision both methods are highly competitive with respect to the best performing conventional and learning-based approaches across all four tasks - with a particular edge for datasets with limited annotations.

2 Methods

In the subsequent two section, we present a brief overview of the methodological principals and algorithmic building blocks of **deeds** and **PDD-net** with a focus on their common and distinct features.

N. Shusharina et al. (Eds.): ABCs 2020/L2R 2020/TN-SCUI 2020, LNCS 12587, pp. 68–73, 2021.
https://doi.org/10.1007/978-3-030-71827-5_8

2.1   Dense Displacement Sampling (Deeds) with Discrete Spanning Tree Optimisation

The main idea behind **deeds** is to explore a very large number of degrees of freedom (of the nonlinear deformation) that correspond to potential discretised displacements. In order to obtain a tractable computational complexity the following three approximations to the exact discrete optimisation problem are made. 1) instead of directly solving the deformation estimation with a single quantised warp, several levels of decreasingly coarse control point grids are employed, 2) patch-based similarity metrics are approximated using a quantised range of values (and the efficient Hamming distance) and a subsampling of voxels, 3) a simplified graph-model, namely a minimum-spanning-tree is employed (that only requires a single forward and backward path of messages for optimisation) in combination with a symmetric inverse consistency approach (see details in [7,9]). By employing the robust yet accurate hand-crafted MIND self-similarity context (SSC) descriptors, the method is applicable to multimodal tasks as well as challenging image appearance, e.g. due to respiration in lung CT or varying contrast in abdominal CT. **deeds** and its extensions (e.g. keypoint matching in [11]) have excelled in lung registration (first place at EMPIRE10 and for the DIR-lab COPD dataset), abdominal registration (MICCAI 2015 beyond the cranial vault challenge) and MRI-US fusion (CuRIOUS challenge MICCAI 2018,2019).

2.2   Probabilistic Dense Displacement (PDD) Net

The **PDD-net** aims to mimic the successful discrete components of **deeds**, while reducing the run-time through GPU implementation by over an order of magnitude and enabling unsupervised feature learning. We firstly use a compact deformable convolutional network [8] to extract features and compute a dense 6D (3 spatial + 3 displacement dimensions) dissimilarity tensor. Employing the minimum-spanning-tree optimisation of **deeds** within a deep learning framework would require the back propagation through dozens of sequential message passing steps. We therefore opted for a simpler graphical model, the mean field inference for a conditional random field (different to Markov random fields no directed messages have to be computed), which enables message passing by (Gaussian) filtering operations over the three spatial dimensions. The label compatibility function (see [2,10] for details) is defined as approximated min-convolutions: a combination of min-filtering and smoothing that act on the three displacement dimensions. These steps are interleaved and repeated twice to predict a discrete probabilistic (softmax) displacement map for each control point. As described in [5] these probabilities can be effectively employed for unsupervised (metric) learning using the aforementioned MIND-SSC descriptors. In most scenarios a single warp of this approach yields very accurate results in less than a second that can be further finetuned using continuous instance optimisation or a second warp.

Table 1. Scores and ranks for each task and the overall L2R Challenge.

	Task1		Task2		Task3		Task4		Overall	
	Score	Rank	Score	Rank	Score	Rank	Score	Rank	Score	Rank
LapIRN	0.72	3	0.75	3	0.93	1	0.95	1	0.83	1
PDD	0.94	1	0.85	1	0.69	4	0.74	4	0.80	2
Deeds	0.91	2	0.79	2	0.73	2	0.55	6	0.73	3
LibReg	0.48	4	0.71	5	0.50	7	0.64	5	0.57	4
Uppsala	0.45	5	0.54	6	0.60	6	0.45	7	0.51	5
CentraleSupélec	0.39	6	0.25	8	0.73	2	0.77	3	0.49	6
Nifty	0.39	6	0.72	4	0.35	8	0.43	8	0.45	7
AGH	0.39	6	0.25	8	0.21	9	0.78	2	0.36	8
KCL	0.39	6	0.25	8	0.68	5	0.20	9	0.34	9
EC Nantes	0.39	6	0.51	7	0.21	9	0.20	9	0.30	10

3   Experiments and Results

For each of the four distinct Learn2Reg tasks we explain the experimental choices
and important hyperparameters. Subsequently, we report numerical results for
each task. Quantitative challenge results can be found in Table 1. The table
shows scores and ranks for each task as well as for the entire challenge. The
scores represent a combined normalized evaluation criterion that includes metrics such as Dice similarity, target registration error (TRE), smoothness of the
transformation (standard deviation of log Jacobian) and runtime of the algorithm. Final ranks are based on significance of scores. Further numerical results
can be found at the Learn2Reg website[1]. Figure 1 shows examples of registration
results of our contributions and selected comparison methods for each task.

Task 1 CuRIOUS US/MRI. This task aims to correct brain shift in multimodal
US/MRI images. **deeds** is used with standard settings for affine transformation
as described in the publicly available code repository[2]. For the **PDD-net** we
chose to use handcrafted MIND features, which showed convincing results for
the multi-modal image modalities in this task. The task poses the problem of
finding an affine transformation between the US and the MRI image. We therefore, estimate an affine matrix from predicted displacements using a trimmed
least square approach. For a robust registration it was also necessary to mask
the ultrasound image. The registration accuracy is evaluated using manual annotated landmarks. For deeds and the PDD-net the TRE is 3.89 mm and 3.09 mm,
respectively. With 4.61 s the PDD-net is almost twice as fast as deeds (9.12 s).

Task 2 Lung CT. The second task deals with aligning inhale and exhale lung
CT scans. We use default settings (See footnote 2) for the **deeds** framework

---

[1] https://learn2reg.grand-challenge.org.
[2] https://github.com/mattiaspaul/deedsBCV.

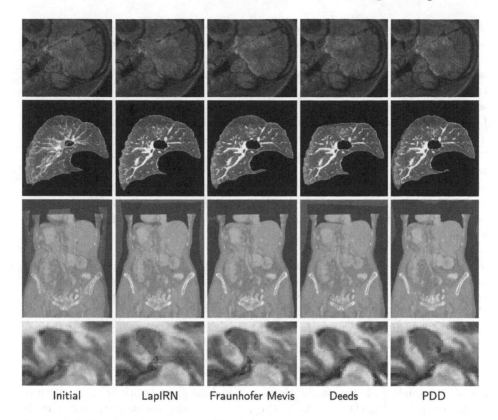

Fig. 1. Qualititative results of contributions and selected comparison methods for all four tasks of the L2R Challenge.

and additionally mask the lung regions. As in the first task we use fixed MIND features for the **PDD-Net**. To exploit the importance of lung vessels for this registration task we implement a sparse variant, where instead of relying on a fixed grid for similarity computations, sparse keypoints are extracted from the inhale scan using the Foerstner operator [3]. To deal with the sparse keypoints within the PDD framework the Gaussian filtering on the fixed grid is replaced by Laplacian smoothing on the kNN graph ($k = 10$). Experiments showed that employing a second warp (after transformation of the moving image) yields significant better results. We therefore employ two warps with the PDD-net using 1024 keypoints in the first and 1536 keypoints in the second warp. As in the first task the registration accuracy is evaluated on manual annotated corresponding landmarks in the inhale and exhale scan. With TREs of 2.26 mm and 2.46 mm the results of deeds and the PDD-net are comparable. However, the runtime of the PDD-net is much faster (2.49 s vs. 41.32 s). The best registration accuracy is achieved by the contribution of Fraunhofer MEVIS with 1.72 mm.

Task 3 Abdominal CT. This tasks aims at the inter-patient registration of abdominal CT scans. Default settings (see Footnote 2) for **deeds** showed convincing results. For the **PDD** registration framework an Obelisk network [8] was used to extract features from the fixed and moving image. The network was trained end-to-end using an unsupervised non-local MIND loss. We employed a second warp after the first transformation of the moving image (using the same feature extraction network). A large boost in registration accuracy with only a small decrease in runtime is observed when using instance optimisation (100 iterations using Adam optimizer) after the initial prediction of displacements with the PDD-net. The registration accuracy is assessed by the Dice similarity of organ segmentation labels. deeds and PDD-net achieve Dice scores of 0.46 and 0.51, while the runtimes are 41.65 s and 4.05 *s*, respectively. The best registration accuracy of all contestants is achieved by the LapIRN framework with a Dice score of 0.67 (using label supervision).

Task 4 Hippocampus MRI. The final task is to align the hippocampus head and body for inter-patient comparison. As the hippocampus MRI images are relatively small ($64 \times 64 \times 64$) we needed to adjust the default parameters (see Footnote 2) for **deeds**. The grid point spacing for the five resolution levels are set to 6,5,4,3 and 2 respectively. For this task the **PDD-net** is extended by a Voxelmorph framework [1]. The combination of both frameworks gave the best registration results in our experiments. For the first warp, the PDD-net with fixed MIND features is used to predict larger deformations. Then, to cope with remaining small scale transformations, a Voxelmorph network is trained using an unsupervised MIND loss. As in the third task the registration accuracy is evaluated using Dice similarity of segmentation labels (hippocampus head and body). For deeds and the PDD-net the Dice score is 0.76 and 0.78, respectively. The runtime for the PDD-net is 10x faster than for deeds (0.31 s vs. 3.14 s). LapIRN achieves the highest Dice score (0.86) using label supervision.

4   Conclusion

In our contributions to the Learn2Reg challenge we analysed the use of two discrete registration methods (**deeds, PDD-net**) with several experimental design choices (MIND features, Obelisk features, instance optimisation, etc.) for the four distinctive challenge tasks. Both proposed methods were ranked among the top three of all challenge contestants, which establishes them as general registration frameworks. The **PDD-net** stands out with fast runtimes and winning task 1 and 2 of the challenge, while **deeds** achieved very consistent registration scores (ranked second for Task 1, 2 and 3).

References

1. Balakrishnan, G., Zhao, A., Sabuncu, M.R., Guttag, J., Dalca, A.V.: Voxelmorph: a learning framework for deformable medical image registration. IEEE Trans. Med. Imaging **38**(8), 1788–1800 (2019)

2. Felzenszwalb, P.F., Huttenlocher, D.P.: Distance transforms of sampled functions. Theory Comput. **8**(1), 415–428 (2012)
3. Förstner, W., Gülch, E.: A fast operator for detection and precise location of distinct points, corners and centres of circular features. In: Intercommission Conference on Fast Processing of Photogrammetric Data, pp. 281–305 (1987)
4. Heinrich, M.P.: Closing the gap between deep and conventional image registration using probabilistic dense displacement networks. In: Shen, D., et al. (eds.) MICCAI 2019. LNCS, vol. 11769, pp. 50–58. Springer, Cham (2019). https://doi.org/10.1007/978-3-030-32226-7_6
5. Heinrich, M.P., Hansen, L.: Highly accurate and memory efficient unsupervised learning-based discrete CT registration using 2.5D displacement search. In: Martel, A.L., et al. (eds.) MICCAI 2020. LNCS, vol. 12263, pp. 190–200. Springer, Cham (2020). https://doi.org/10.1007/978-3-030-59716-0_19
6. Heinrich, M.P., et al.: Mind: Modality independent neighbourhood descriptor for multi-modal deformable registration. Med. Image Anal. **16**(7), 1423–1435 (2012)
7. Heinrich, M.P., Jenkinson, M., Brady, S.M., Schnabel, J.A.: MRF-based deformable registration and ventilation estimation of lung CT. IEEE Trans. Med. Imaging **32**(7), 1239–48 (2013)
8. Heinrich, M.P., Oktay, O., Bouteldja, N.: Obelisk-net: fewer layers to solve 3D multi-organ segmentation with sparse deformable convolutions. Med. Image Anal. **54**, 1–9 (2019)
9. Heinrich, M.P., Jenkinson, M., Papież, B.W., Brady, S.M., Schnabel, J.A.: Towards realtime multimodal fusion for image-guided interventions using self-similarities. In: Mori, K., Sakuma, I., Sato, Y., Barillot, C., Navab, N. (eds.) MICCAI 2013. LNCS, vol. 8149, pp. 187–194. Springer, Heidelberg (2013). https://doi.org/10.1007/978-3-642-40811-3_24
10. Krähenbühl, P., Koltun, V.: Efficient inference in fully connected CRFs with Gaussian edge potentials. Adv. Neural Inf. Process. Syst. **24**, 109–117 (2011)
11. Rühaak, J., et al.: Estimation of large motion in lung CT by integrating regularized keypoint correspondences into dense deformable registration. IEEE Trans. Med. Imaging **36**(8), 1746–1757 (2017)

Variable Fraunhofer MEVIS RegLib Comprehensively Applied to Learn2Reg Challenge

Stephanie Häger⬥, Stefan Heldmann⬥, Alessa Hering(✉)⬥, Sven Kuckertz⬥, and Annkristin Lange⬥

Fraunhofer Institute for Digital Medicine MEVIS, Lübeck, Germany
`alessa.hering@mevis.fraunhofer.de`

Abstract. In this paper, we present our contribution to the learn2reg challenge. We applied the Fraunhofer MEVIS registration library RegLib comprehensively to all 4 tasks of the challenge. For tasks 1–3, we used a classic iterative registration method with NGF distance measure, second order curvature regularizer, and a multi-level optimization scheme. For task 4, a deep learning approach with a weakly supervised trained U-Net was applied using the same cost function as in the iterative approach.

Keywords: Image registration · Registration challenge · Learn2Reg

1 Introduction

Image registration is a key task in medical image analysis to estimate deformations between images and to obtain spatial correspondences. The goal of image registration is to find a reasonable deformation for a pair of fixed and moving image so that the transformed moving image and the fixed image are similar. Image registration is typically formulated as an optimization problem where a suitable cost function is minimized through iterative optimization schemes. Over time, a variety of image registration models and approaches have been developed. Therefore comparison possibilities are needed. In order to ensure comparability, challenges are created in which the different registration procedures are evaluated on the same image data and under the same computation conditions. One such challenge is the *Learn2Reg: 2020 MICCAI Registration Challenge* [1,3]. It consists of 4 different registration tasks that cover both intra- and inter-patient alignment, CT, ultrasound and MRI modalities, neuro-, thorax and abdominal anatomies. In this paper we present our solutions to all 4 tasks of the challenge.

2 Method and Results

Task 1–3 is solved by classical iterative methods. Task 4 is tackled by an U-Net that has been weakly supervised trained for end-to-end registration. All tasks

© Springer Nature Switzerland AG 2021
N. Shusharina et al. (Eds.): ABCs 2020/L2R 2020/TN-SCUI 2020, LNCS 12587, pp. 74–79, 2021.
https://doi.org/10.1007/978-3-030-71827-5_9

build on cost functions and losses made up from several terms that are selected for the specific task. Common to all is the use of normalized gradient fields (NGF) [5] image similarity for fixed and moving images $\mathcal{F}, \mathcal{M} : \Omega \subset \mathbb{R}^3 \rightarrow \mathbb{R}$

$$\mathrm{NGF}(\mathcal{F}, \mathcal{M}) = \frac{1}{2} \int_\Omega 1 - \frac{\langle \nabla\mathcal{F}, \nabla\mathcal{M} \rangle^2_{\epsilon_\mathcal{F}\epsilon_\mathcal{M}}}{\|\nabla\mathcal{F}\|^2_{\epsilon_\mathcal{F}} \|\nabla\mathcal{M}\|^2_{\epsilon_\mathcal{M}}} \, dx \tag{1}$$

with parameters $\epsilon_\mathcal{F}, \epsilon_\mathcal{M} > 0$, $\langle x, y \rangle_\epsilon := x^\mathsf{T} y + \epsilon$ and $\|x\|_\epsilon = \sqrt{\langle x, y \rangle_\epsilon}$ and 2nd order curvature (CURV) regularization [4] of displacement vector fields $u : \Omega \subset \mathbb{R}^3 \rightarrow \mathbb{R}^3$

$$\mathrm{CURV}(u) = \frac{1}{2} \int_\Omega \sum_{\ell=1}^{3} \|\Delta u_\ell\|^2 \, dx. \tag{2}$$

Furthermore, the methods for task 1–3 use a coarse-to-fine multi-level iterative registration scheme where a Gaussian image pyramid is generated for both images to obtain downsampled and smoothed images. Then, a registration is performed on the lowest resolution level and the resulting deformation field serves as the starting point for the following registration on the next highest level. This proceeds till the finest level with quasi-Newton L-BFGS optimization at each level. This procedure allows to align larger structures on the lower levels and helps to avoid local minima, to reduce topological changes or foldings, and to speed up run times.

Metrics for accuracy (TRE, DICE, Hausdorff95), robustness (DICE30) and plausibility of the deformation field (LogJacDetStd) are computed for evaluation of the challenge. More details can be found in [2]. Table 1 shows the results of our methods for all 4 tasks. The runtimes for our methods were not measured during the challenge evaluation itself, but measured afterwards performing our methods on the challenge data with a NVIDIA GeForce RTX 2070 with 8 GB memory and an Intel Core i7-9700K.

Task 1
The first task deals with the challenge of multimodal MRI vs. US registration of the brain. The provided data was preprocessed including a resampling to the size of $256 \times 256 \times 288$ voxels at an isotropic 0.5 mm resolution. We focused on a parametric rigid registration due to the fact that we assume only minor local deformations inside the brain. The registration consists of two stages: an initial fast translational alignment and a multi-level registration restricted to rigid deformations. In both stages, the NGF distance measure with parameters $\epsilon_F = 3$ and $\epsilon_M = 2$ is minimized. In contrast to the task definition, we use the US images as fixed and the FLAIR MRIs as moving images, respectively. We changed the roles of the images in order to mask the distance measure to the significally smaller US volume using a threshold. Afterwards, the resulting registration matrix is inverted and transformed to a corresponding displacement field. While our method successfully accomplishes all registrations on the validation set (improving the TRE from 7.02 mm to 2.75 mm on average for 7 cases), we had difficulties in some cases of the test set. This can be seen in Table 1, where a

Table 1. Results of our methods in the Learn2Reg Challenge. For the challenge the target registration error (TRE), DICE score, robustness score (30% lowest DICE of all cases, DICE30), 95% percentile of the Hausdorff distance (Hausdorff95) and the standard deviation of log Jacobian determinant of the deformation field (LogJacDetStd) were measured. We additionally measured the runtimes for our methods on a local machine.

	Task 1	Task 2	Task 3	Task 4
TRE before reg. [mm]	6.37 ± 4.34	10.24 ± 6.57	-	-
TRE after reg. [mm]	7.02 ± 5.95	1.72 ± 0.59	-	-
DICE before reg.	-	-	0.23 ± 0.21	0.55 ± 0.15
DICE after reg.	-	-	0.47 ± 0.29	0.85 ± 0.04
DICE30 before reg.	-	-	0.01	0.36
DICE30 after reg.	-	-	0.21	0.84
Hausdorff95 before reg. [mm]	-	-	46.07	3.91
Hausdorff95 after reg. [mm]	-	-	43.32	1.55
LogJacDetStd	0.00	0.07	0.14	0.05
runtime [s]	14.74 ± 5.84	92.71 ± 17.22	0.97 ± 0.23	0.39 ± 0.13

slightly increasing TRE is shown. However, in 5 out of 10 test cases we improve the TRE from 5.84 mm to 2.86 mm on average. The partial inexact registrations could be the result of the difficult parameterization and therefore generalization of the challenging task of MRI vs. US registration and need further investigation.

Task 2

The aim of the second task was the registration of expiration to inspiration CT scans of the lung. The provided data consists of 20 training scan pairs [8] and 10 test scan pairs [7]. All scan pairs were resampled to a image size of $192 \times 192 \times 208$ and were affine pre-registered. The main challenges are the large deformation due to breathing and that the lungs in the expiration scans are not fully visible.

Our submitted method based on our previous work [10]. First, a graph-based matching of a large number of keypoints for the estimation of robust large-motion correspondences is performed. Then, this is followed by a continuous, deformable image registration incorporating both image intensities and keypoint information. Herefore, we used the NGF distance measure with edge parameter $\epsilon = 0.1$. For a smooth deformation field we selected the curvature regularizer with weight parameter $\alpha = 1$. In contrast to [10], we are not integrating the lung mask into a cost term to enforce lung boundary alignment, because the expiration lung is not fully visible. However, we mask the NGF distance measure with the expiration lung mask. A coarse-to-fine multilevel scheme with 3 levels was applied. With a target registration error of 1.72 ± 0.59 mm, we archived the highest accuracy of all submissions in the challenge. The whole registration pipeline takes about 92.7 s which includes the keypoint detection with 86.4 s

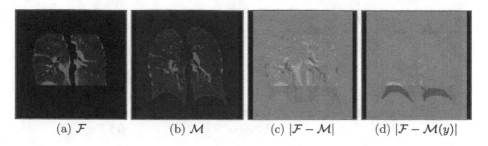

(a) $\mathcal{F}$ (b) $\mathcal{M}$ (c) $|\mathcal{F} - \mathcal{M}|$ (d) $|\mathcal{F} - \mathcal{M}(y)|$

Fig. 1. Example coronal slices extracted from a exemplary case for task 2: a) The expiration image, b) inspiration image, c) the difference image before the registration and d) the difference image after registration. For a better visualization, we only show the image inside the lung, however, the full thorax scan was used.

and the actual registration with 6 s. All results are summarized in Table 1. To illustrate the registration results, we show the difference images $\mathcal{F} - \mathcal{M}(y)$ before and after registration in Fig. 1. The breathing motion was successfully recovered and inner lung structures are well aligned.

Task 3

The aim of the third task was the inter-patient registration of abdominal CT scans to transfer organ segmentations. The 30 training and 20 test CT scans were provided with preprocessing such as same isotropic voxel resolutions (2 mm) and spatial dimensions ($192 \times 160 \times 256$) as well as affine preregistration. Therefore we only use a non-parametric registration approach with the normalized gradient fields as a distance measure. The edge parameter ϵ was set to $\epsilon_F = \epsilon_M = 5$. To achieve a smooth deformation field we selected the curvature regularizer weighted with parameter $\alpha = 10$. A multilevel scheme with 4 levels was applied, where the resolution of the deformation field was one level lower than the image resolution. For all specified parameters different values were tested and the ones chosen that led to the best results regarding the DICE score on the training images. To solve the optimization problem efficiently, we used an implementation on the graphics card. This reduced the average runtime for a single registration from 25 s to less than 1 s. As shown in Table 1 the DICE score could be improved from 0.23 to 0.47 on the test CT scans. In Fig. 2 an exemplary registration result on the validation data with the overlayed segmentations is shown. The example illustrates the difficulty of interpatient registration due to the large anatomical differences. Partially the registration can compensate for this, but especially if the anatomical differences are large but the intensity changes are small, a good registration result is difficult to achieve.

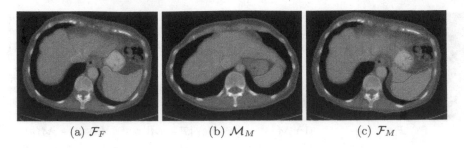

(a) $\mathcal{F}_F$ (b) $\mathcal{M}_M$ (c) $\mathcal{F}_M$

Fig. 2. Exemplary registration result for task 3. The original fixed and moving images are shown in axial direction with their according labels (a, b). In (c) the fixed CT is overlayed by the deformed moving labels.

Task 4

The data for the fourth task consists of 394 MRI scans covering the hippocampus formation in the brain, divided into 263 and 131 scans for training and testing, respectively. The aim of this task is the alignment of the hippocampus head and body using inter-patient registration. For the training cases segmentations of these two structures are available and the data is additionally provided preprocessed to same voxel resolutions and spatial dimensions. In order to accomplish this task, we train a U-Net with four levels in weakly supervised manner, aiming to optimize the NGF image similarity with $\epsilon_F = \epsilon_M = 1$ and a curvature deformation regularizer including a penalty for implausible grid foldings. Additionally, we measure the alignment of the given segmentations using a sum of squared differences [6,9]. Our method receives only a fixed and a moving image as input, computes the displacement field at the same resolution and includes corresponding segmentations exclusively during training (weak supervision). Furthermore, our training is not depending on difficult to access ground-truth displacement fields. After the training of our network, only a single pass through the network is required for registration of unseen image pairs, computing displacement fields in a fraction of a second (on a GPU). As shown in Table 1, the weakly supervised training enables our method to improve the DICE score on average from 0.55 to 0.85, while maintaining physically plausible results.

3   Conclusion

We showed that the Fraunhofer MEVIS RegLib is successfully applicable to all 4 tasks of the Learn2Reg challenge that differ greatly and cover both intra- and inter-patient alignment, various modalities and anatomies. We chose the iterative method for tasks 1–3 due to the limited amount of data available and a deep learning approach for task 4. Our methods achieved the forth place in the challenge without consideration of our fast runtime. Furthermore, we achieved the overall highest registration accuracy with our method in task 2.

References

1. Learn2Reg: 2020 MICCAI registration challenge. https://learn2reg.grand-challenge.org/
2. Learn2Reg challenge, metrics and evaluation. https://learn2reg.grand-challenge.org/Submission/
3. Dalca, A., et al.: Learn2Reg - the challenge, March 2020. https://doi.org/10.5281/zenodo.3715652
4. Fischer, B., Modersitzki, J.: Curvature based image registration. J. Math. Imaging Vis. **18**(1), 81–85 (2003)
5. Haber, E., Modersitzki, J.: Intensity gradient based registration and fusion of multi-modal images. In: Medical Image Computing and Computer-Assisted Intervention - MICCAI 2006, vol. 3216, pp. 591–598 (2006)
6. Hering, A., Kuckertz, S., Heldmann, S., Heinrich, M.P.: Enhancing label-driven deep deformable image registration with local distance metrics for state-of-the-art cardiac motion tracking. Bildverarbeitung für die Medizin 2019. I, pp. 309–314. Springer, Wiesbaden (2019). https://doi.org/10.1007/978-3-658-25326-4_69
7. Hering, A., Murphy, K., van Ginneken, B.: Learn2Reg Challenge: CT Lung Registration - Test Data, September 2020. https://doi.org/10.5281/zenodo.4048761
8. Hering, A., Murphy, K., van Ginneken, B.: Learn2Reg Challenge: CT Lung Registration - Training Data, May 2020. https://doi.org/10.5281/zenodo.3835682
9. Kuckertz, S., Papenberg, N., Honegger, J., Morgas, T., Haas, B., Heldmann, S.: Deep learning based CT-CBCT image registration for adaptive radio therapy. In: Medical Imaging 2020: Image Processing, vol. 11313, pp. 149–154. International Society for Optics and Photonics, SPIE (2020)
10. Rühaak, J., Polzin, T., Heldmann, S., Simpson, I.J., Handels, H., Modersitzki, J., Heinrich, M.P.: Estimation of large motion in lung CT by integrating regularized keypoint correspondences into dense deformable registration. IEEE Trans. Med. Imaging **36**(8), 1746–1757 (2017)

Learning a Deformable Registration Pyramid

Niklas Gunnarsson$^{1,2(\boxtimes)}$ (ID), Jens Sjölund$^{1,2(\boxtimes)}$ (ID), and Thomas B. Schön$^{1(\boxtimes)}$ (ID)

^{1} Department of Information Technology, Uppsala University, Uppsala, Sweden
{niklas.gunnarsson,jens.sjolund,thomas.schon}@it.uu.se
^{2} Elekta Instrument AB, Stockholm, Sweden
{niklas.gunnarsson,jens.sjolund}@elekta.com

Abstract. We introduce an end-to-end unsupervised (or weakly supervised) image registration method that blends conventional medical image registration with contemporary deep learning techniques from computer vision. Our method downsamples both the fixed and the moving images into multiple feature map levels where a displacement field is estimated at each level and then further refined throughout the network. We train and test our model on three different datasets. In comparison with the initial registrations we find an improved performance using our model, yet we expect it would improve further if the model was fine-tuned for each task. The implementation is publicly available (https://github.com/ngunnar/learning-a-deformable-registration-pyramid).

Keywords: Medical image registration · Deep learning · Deformable registration.

1 Introduction

Image registration is a fundamental problem in medical imaging. It is widely used in applications to, for example, combine images of the same object from different modalities (multimodal registration), detect changes between images at different times (spatiotemporal registration), and map segments from a predefined image to a new image (atlas based segmentation).

The basic principle of image registration is to find a displacement field ϕ that maps positions in a moving image to the corresponding positions in a fixed image. Conventionally, image registration problems are often stated as optimization problems, where the aim is to minimize a complex energy function [1].

A popular heuristic for solving image registration problems is to use a coarse-to-fine approach [2] i.e. to start with a rough estimate of the displacement field and refine it in one or several steps. It is common to downsample the fixed and moving images using a kernel based pyramid, and make a first estimate of the displacement field at the lowest resolution which is then used as an initial guess when estimating the field at the next resolution level, and so forth.

© Springer Nature Switzerland AG 2021
N. Shusharina et al. (Eds.): ABCs 2020/L2R 2020/TN-SCUI 2020, LNCS 12587, pp. 80–86, 2021.
https://doi.org/10.1007/978-3-030-71827-5_10

Due to the complexity of the energy function each estimate is computationally expensive and requires long execution time. Machine learning provides an alternative approach, where a model is optimized (learned) offline based on a training dataset, obviating the need for expensive optimization at test time [3]. In this paper we present an image registration method that combines the conventional coarse-to-fine approach with a convolutional neural network (CNN).

2   Method

We have developed a 3D deformable image registration method inspired by the PWC-Net [4], a 2D optical flow method popular in computer vision. Our method estimates and refines a displacement field at each level of a CNN downsampling pyramid.

2.1   Architecture

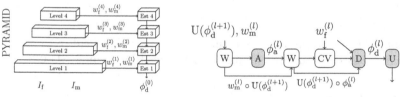

(a) Model architecture.   (b) Operations at each feature level.

Fig. 1. An overview of the model architecture. The moving and fixed image are downsampled into several feature maps using the pyramid (a). Figure (b) shows operations at each feature level. Blue and white boxes represent operations with and without trainable parameters, respectively.

The pyramid downsamples the moving image I_{m} and the fixed image I_{f} into several feature maps $\{w_{\mathrm{m}}^{(l)}, w_{\mathrm{f}}^{(l)}\}_{l=1}^{L}$. At each level, starting from the top, a displacement field ϕ_{d}^{L} is estimated and used as an initial guess at finer levels. Figure 1 illustrates the model architecture (a) and operations at each level (b). The total number of trainable parameters in our model is 8.6 million. Our model includes multiple CNN blocks. These consist of a 3D convolutional layer followed by Leaky Rely and batch normalization. All 3D convolutional layers use a kernel size of (3,3,3). Each module of our model is explained below:

Pyramid: Downsamples the moving and fixed image into several feature map levels using 3D CNN layers. The same pyramid is used for the moving and the fixed images. We use a four-level pyramid ($L = 4$) where each level consists of three CNN blocks. The stride is two in the first block and one in the subsequent blocks. The number of filters at each level is 16, 32, 32, and 32, respectively.

Warp (W): Warps features from moving images with the estimated displacement field. This module has no trainable parameters.

Affine (A): A dense neural network that estimates the 12 parameters in an affine transformation. This module consists of a global average pooling followed by a dense layer.

Cost volume (CV): Correlation between the warped feature maps from the moving image and feature maps from the fixed image. For computational reasons the cost volume is restricted to voxel neighborhoods of size d. This module has no trainable parameters.

Deform (D): A 3D DenseNet [5] that estimates the displacement field based on its current estimate, the cost volume and the feature maps from the fixed image. This module uses 5 CNN blocks of the same type as in the Pyramid but with 64, 64, 32, 18, and 8 filters, respectively followed by a convolutional layer with 3 filters.

Upsample (U): Upsamples the estimated displacement field from one level to the next. Consists of an upsampling layer followed by a single 3D CNN.

2.2 Loss Function

Our loss function combines image similarity with regularization of the displacement field. By including the intermediate estimates in the loss, we aim to gain additional control of the network. Auxiliary information, e.g. anatomical segmentations S_m and S_f are incorporated via an additional structural similarity term $\mathcal{L}_{\text{seg}}$. Our resulting loss function can be written as

$$\mathcal{L} = \mathcal{L}_{\text{seg}} + \sum_{l=0}^{L} \left(\mathcal{L}_{\text{sim}}^{(l)} + \mathcal{L}_{\text{smooth}}^{(l)} \right). \tag{1}$$

We use the (soft) Dice coefficient (DCS) [6] for structural similarity and the normalized cross-correlation (NCC) [7] for image similarity. To ensure smooth displacements we regularize the affine displacement field with the L2-loss between the estimated value and an identity displacement field ($\phi_0^{(l)}$) and the deformable field with the spatial gradient of the displacement field [8],

$$\mathcal{L}_{\text{seg}} \left(S_f, S_m, \phi_d^{(0)} \right) = \lambda(1 - \text{DCS}(S_f, S_m \circ \phi_d^{(0)})), \tag{2a}$$

$$\mathcal{L}_{\text{sim}}^{(l)} \left(I_f^{(l)}, I_m^{(l)}, \phi_d^{(l)} \right) = -\gamma^{(l)} \text{NCC}(I_f^{(l)}, I_m^{(l)} \circ \phi_d^{(l)}), \tag{2b}$$

$$\mathcal{L}_{\text{smooth}}^{(l)} \left(\phi_a^{(l)}, \phi_d^{(l)} \right) = \alpha^{(l)} ||\phi_a^{(l)} - \phi_0^{(l)}||_2^2 + \beta^{(l)} ||\nabla \phi_d^{(l)}||_2^2, \tag{2c}$$

where $I_m^{(l)}$ and $I_f^{(l)}$ represent downsampled versions of the moving and fixed images at each level and $\phi_a^{(l)}$ and $\phi_d^{(l)}$ indicate the estimated affine and deformable registrations (for each level). The hyperparameters $\lambda, \{\gamma^{(l)}, \alpha^{(l)} \text{ and } \beta^{(l)}\}_{l=0}^{L}$ determine the importance of the corresponding terms.

3   Experiment

We evaluated the model on three different tasks from the 2020 Learn2Reg challenge [9]. The different tasks were: inspiration and expiration CT scans of thorax images with automatic segmented lung (Task 2) [10]; 3D CT abdominal images with thirteen segmented organs (Task 3); and segmented hippocampus MRI of healthy adults and adults with non-affective psychotic disorder (Task 4) [11].

We trained our model on image pairs from all tasks at the same time. All images were downsampled (to a resolution of $64 \times 64 \times 64$) and normalized ($I_f, I_m \in [0, 1]$). The different hyperparameters were $\lambda = 5.0$, $\gamma^{(l)} = 5/2^l$, $\alpha^{(l)} = 2^l$ and $\beta^{(l)} = 1/2^l$ for $l \in \{0, \ldots, 4\}$ and for cost volume search range we used $d = 2$. The network was trained end-to-end using the Adam optimizer and a learning rate of 10^{-4}. To speed up training we used distributed training on three Nvidia GeForce GTX 1080 Ti graphic cards and trained the model for 100 epochs, which took approximately 24 h.

The results are shown in Table 1. Table 2 shows examples of warping the moving image using displacement fields $\phi_d^{(l)}$ estimated at three different levels $l \in \{0, 2, 4\}$. Based on the total score, our approach was ranked 5th according to the public leaderboard [9].

Table 1. Result on test dataset for each task.

Task	Method	TRE [12]	TRE30	DCS [13]	DCS30	HD95 [14]	SDlogJ [15]	Time (s) GPU	Time (s) CPU[a]
2	Our	9.00	12.22	–	–	–	0.12	0.31	4.83
	Initial	10.24	17.77	–	–	–	0.00	–	–
3	Our	–	–	0.39	0.12	43.03	0.13	0.31	4.83
	Initial	–	–	0.23	0.01	46.07	0.00	–	–
4	Our	–	–	0.74	0.67	2.82	0.16	0.32	4.83
	Initial	–	–	0.55	0.36	3.91	0.00	–	–

[a] Prediction time only, excluding pre - and post processing.

Table 2. Sample result from the validation dataset. The moving image I_m (left) is warped with the estimated displacement field from several levels ($l = 4, 2, 0$). Starting from the coarsest to the finest level. The fixed image I_f is shown to the right.

I_m	$I_m \circ \phi_d^{(4)}$	$I_m \circ \phi_d^{(2)}$	$I_m \circ \phi_d^{(0)}$	I_f

4   Conclusion and Future Work

In this paper we have shown that it is possible to include domain knowledge when developing machine learning methods for medical image registration problems. Our method operates in a coarse-to-fine manner and could be modified in many ways, e.g. by replacing the CNN pyramid with other technologies; like a Laplacian pyramid, similar to the winner of the competition [16], or modifying/removing displacement fields estimations (affine or deformable) in the levels.

In comparison with other participants in the competition our approach was to create a single general model for all tasks while other participants used different models or different training procedures [16–18] for each task. The general approach showed increased performance compared with initial registrations. In future work, we will evaluate to what extent the performance improves when fine tuning the model for each task.

During the training phase the memory usage was high (11.4 GB). In the experiments we downsampled the input images to a low resolution, using a batch size of one (at each GPU replica) and our partial cost volume had a search range of two to be able to fit the model in GPU memory (11.7 GB). We believe that an in-depth analysis of the network will reveal ways of reducing memory usage without sacrificing performance substantially, e.g. by removing superfluous layers or reducing the number of filters. One idea is to reduce the number of parameters in the DenseNet [19]. Other potential improvements include: 1) training each level separately, starting from the coarsest, which will reduce the number of trainable parameters in each training process, 2) training the model on slices (2D) or thin slabs (2.5D), instead of the entire volume and iteratively estimate the entire 3D displacement field.

Acknowledgement. This research was funded by the *Wallenberg AI, Autonomous Systems and Software Program (WASP)* funded by Knut and Alice Wallenberg Foundation, and the Swedish Foundation for Strategic Research grant SM19-0029.

References

1. Hill, D.L.G., Batchelor, P.G., Holden, M., Hawkes, D.J.: Medical image registration. Phys. Med. Biol. **46**(3), R1 (2001)
2. Zitova, B., Flusser, J.: Image registration methods: a survey. Image Vis. Comput. **21**(11), 977–1000 (2003)
3. Balakrishnan, G., Zhao, A., Sabuncu, M.R., Guttag, J., Dalca, A.V.: VoxelMorph: a learning framework for deformable medical image registration. IEEE Trans. Med. Imaging **38**(8), 1788–1800 (2019)
4. Sun, D., Yang, X., Liu, M.Y., Kautz, J.: PWC-Net: CNNs for optical flow using pyramid, warping, and cost volume. In Proceedings of the IEEE Conference on Computer Vision and Pattern Recognition, pp. 8934–8943 (2018)
5. Huang, G., Liu, Z., Van Der Maaten, L., Weinberger, K.Q.: Densely connected convolutional networks. In: Proceedings of the IEEE Conference on Computer Vision and Pattern Recognition, pp. 4700–4708 (2017)
6. Milletari, F., Navab, N., Ahmadi, S.A.: V-Net: fully convolutional neural networks for volumetric medical image segmentation. In: 2016 Fourth International Conference on 3D Vision (3DV), pp. 565–571. IEEE (2016)
7. Pratt, W.K.: Digital image processing, 4th edition. J. Electron. Imaging **16**(2), 29901 (2007)
8. Estienne, T., et al.: U-ReSNet: ultimate coupling of registration and segmentation with deep nets. In: Shen, D., et al. (eds.) MICCAI 2019. LNCS, vol. 11766, pp. 310–319. Springer, Cham (2019). https://doi.org/10.1007/978-3-030-32248-9_35
9. Dalca, A., et al.: Learn2reg - the challenge (2020). https://learn2reg.grand-challenge.org/, https://doi.org/10.5281/ZENODO.3715652
10. Hering, A., Murphy, K., van Ginneken, B.: Lean2Reg challenge: CT lung registration - training data, May 2020
11. Simpson, A.L., et al.: A large annotated medical image dataset for the development and evaluation of segmentation algorithms. arXiv preprint arXiv:1902.09063 (2019)
12. Fitzpatrick, J.M., West, J.B., Maurer, C.R.: Predicting error in rigid-body point-based registration. IEEE Trans. Med. Imaging **17**(5), 694–702 (1998)
13. Dice, L.R.: Measures of the amount of ecologic association between species. Ecology **26**(3), 297–302 (1945)
14. Huttenlocher, D.P., Klanderman, G.A., Rucklidge, W.J.: Comparing images using the Hausdorff distance. IEEE Trans. Pattern Anal. Mach. Intell. **15**(9), 850–863 (1993)
15. Kabus, S., Klinder, T., Murphy, K., van Ginneken, B., Lorenz, C., Pluim, J.P.W.: Evaluation of 4D-CT lung registration. In: Yang, G.-Z., Hawkes, D., Rueckert, D., Noble, A., Taylor, C. (eds.) MICCAI 2009. LNCS, vol. 5761, pp. 747–754. Springer, Heidelberg (2009). https://doi.org/10.1007/978-3-642-04268-3_92

16. Mok, T.C.W., Chung, A.C.S.: Large deformation diffeomorphic image registration with Laplacian pyramid networks. In: Martel, A.L., et al. (eds.) MICCAI 2020. LNCS, vol. 12263, pp. 211–221. Springer, Cham (2020). https://doi.org/10.1007/978-3-030-59716-0_21

17. Heinrich, M.P.: Closing the gap between deep and conventional image registration using probabilistic dense displacement networks. In: Shen, D., et al. (eds.) MICCAI 2019. LNCS, vol. 11769, pp. 50–58. Springer, Cham (2019). https://doi.org/10.1007/978-3-030-32226-7_6

18. Estienne, T.: Deep learning based registration using spatial gradients and noisy segmentation labels. arXiv preprint arXiv:2010.10897 (2020)

19. Jégou, S., Drozdzal, M., Vazquez, D., Romero, A., Bengio., Y.: The one hundred layers tiramisu: fully convolutional DenseNets for semantic segmentation. In: Proceedings of the IEEE Conference on Computer Vision and Pattern Recognition Workshops, pp. 11–19 (2017)

Deep Learning Based Registration Using Spatial Gradients and Noisy Segmentation Labels

Théo Estienne[1,2(✉)], Maria Vakalopoulou[1], Enzo Battistella[1,2],
Alexandre Carré[2], Théophraste Henry[2], Marvin Lerousseau[1,2],
Charlotte Robert[2], Nikos Paragios[3], and Eric Deutsch[2]

[1] Université Paris-Saclay, CentraleSupélec, Mathématiques et Informatique pour la Complexité et les Systémes, Inria Saclay, 91190 Gif-sur-Yvette, France
`theo.estienne@centralesupelec.fr`
[2] Université Paris-Saclay, Institut Gustave Roussy, Inserm, Radiothérapie Moléculaire et Innovation Thérapeutique, 94800 Villejuif, France
[3] Therapanacea, Paris, France

Abstract. Image registration is one of the most challenging problems in medical image analysis. In the recent years, deep learning based approaches became quite popular, providing fast and performing registration strategies. In this short paper, we summarise our work presented on Learn2Reg challenge 2020. The main contributions of our work rely on *(i)* a symmetric formulation, predicting the transformations from source to target and from target to source simultaneously, enforcing the trained representations to be similar and *(ii)* integration of variety of publicly available datasets used both for pretraining and for augmenting segmentation labels. Our method reports a mean dice of 0.64 for task 3 and 0.85 for task 4 on the test sets, taking third place on the challenge. Our code and models are publicly available at https://github.com/TheoEst/ abdominal_registration and https://github.com/TheoEst/hippocampus_ registration.

1 Introduction

In the medical field, the problem of deformable image registration has been heavily studied for many years. The problem relies on establishing the best dense voxel-wise transformation (Φ) to wrap one volume (source or moving, M) to match another volume (reference or fixed, F) in the best way. Traditionally, different types of formulations and approaches had been proposed in the last years [17] to address the problem. However, with the recent advances of deep learning, a lot of learning based methods became very popular currently, providing very efficient and state-of-the art performances [9]. Even if there is a lot of work in the field of image registration there are still a lot of challenges to be addressed. In order to address these challenges and provide common datasets for the benchmarking of learning based [5,19] and traditional methods [1,10], the

© Springer Nature Switzerland AG 2021
N. Shusharina et al. (Eds.): ABCs 2020/L2R 2020/TN-SCUI 2020, LNCS 12587, pp. 87–93, 2021.
https://doi.org/10.1007/978-3-030-71827-5_11

Learn2Reg challenge is organised [4]. Four tasks were proposed by the organisers with different organs and modalities. In this work, we focused on two tasks: the CT abdominal (task 3) and the MRI hippocampus registration (task 4).

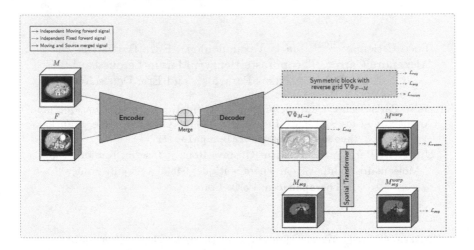

Fig. 1. Schematic representation of the proposed methodology.

In this work, we propose a learning based method that learns how to obtain spatial gradients in a similar way to [6,18]. The main contributions of this work rely on *(i)* enforcing the same network to predict both $\Phi_{M \to F}$ and $\Phi_{F \to M}$ deformations using the same encoding and implicitly enforcing it to be symmetric and *(ii)* integrating noisy labels from different organs during the training, to fully exploit publicly available datasets. In the following sections, we will briefly summarise these two contributions and present our results that gave to our method the third position in the Learn2Reg challenge 2020 (second for task 3 and third for task 4).

2 Methodology

An overview of our proposed framework is presented in the Fig. 1. Our method uses as backbone a 3D UNet [3] based architecture, which consists of 4 blocks with 64, 128, 256 and 512 channels for the encoder part (**E**). Each block consists of a normalisation layer, Leaky ReLU activation, 3D convolutions with a kernel size of $3 \times 3 \times 3$ and convolution with kernel size and stride 2 to reduce spatial resolution. Each of the F, M volumes passes independently through the encoder part of the network. Their encoding is then merged using the subtraction operation before passing through the decoder (**D**) part for the prediction of the optimal spatial gradients of the deformation field $\nabla \Phi$. We obtained the deformation field Φ from its gradient using integration which we approximated with the cumulative summation operation. Φ is then used to obtain the deformed volume together with its segmentation mask using warping $M^{warp} = \mathcal{W}(M, \Phi_{M \to F})$. Finally, we apply deep supervision to train our network in a way similar to [13].

Symmetric Training. Even if our grid formulation has constraints for the spatial gradients to avoid self-crossings on the vertical and horizontal directions for each of the x,y,z-axis, our formulation is not diffeomorphic. This actually indicates that we can not calculate the inverse transformation of $\Phi_{M \to F}$. To deal with this problem, we predict both $\Phi_{M \to F}$ and $\Phi_{F \to M}$ and we use both for the optimization of our network. Different methods such as [8,12] explore similar concepts using however different networks for each deformation. Due to our fusion strategy on the encoding part, our approach is able to learn both transformations with less parameters. In particular, our spatial gradients are obtained by: $\nabla \Phi_{M \to F} = \mathbf{D}(\mathbf{E}(M) - \mathbf{E}(F))$ and $\nabla \Phi_{F \to M} = \mathbf{D}(\mathbf{E}(F) - \mathbf{E}(M))$.

Pretraining and Noisy Labels. Supervision has been proved to boost the performance of the learning based registration methods integrating implicit anatomical knowledge during the training procedure. For this reason, in this study, we investigate ways to use publicly available datasets to boost performance. We exploit available information from publicly available datasets namely KITS 19 [11], Medical Segmentation Decathlon (sub-cohort Liver, Spleen, Pancreas, Hepatic Lesion and Colon) [16] and TCIA Pancreas[7,15]. In particular, we trained a 3D UNet segmentation network on 11 different organs (spleen, right and left kidney, liver, stomach, pancreas, gallbladder, aorta, inferior vena cava, portal vein and oesophagus). To harmonise the information that we had at disposal for each dataset, we optimised the dice loss only on the organs that were available per dataset. The network was then used to provide labels for the 11 organs for approximately 600 abdominal scans. These segmentation masks were further used for the pretraining of our registration network for the task 3. After the training the performance of our segmentation network on the validation set in terms of dice is summarised to: 0.92 (Spl), 0.90 (RKid), 0.91 (LKid), 0.94 (Liv) 0.83 (Sto), 0.74 (Pan), 0.72 (GBla), 0.89 (Aor), 0.76 (InfV), 0.62 (PorV) and 0.61 (Oes). The validation set was composed of 21 patients of Learn2Reg and TCIA Pancreas dataset.

Furthermore, we explored the use of pretraining of registration networks on domain-specific large datasets. In particular, for task 3 the ensemble of the publicly available datasets together with their noisy segmentation masks were used to pretrain our registration network, after a small preprocessing including an affine registration step using Advanced Normalization Tools (ANTs) [2] and isotropic resampling to 2 mm voxel spacing. Moreover, for task 4, we performed an unsupervised pretraining using approximately 750 T1 MRI from OASIS 3 dataset [14] without segmentations. For both tasks, the pretraining had been performed for 300 epochs.

2.1   Training Strategy and Implementation Details

To train our network, we used a combination of multiple loss functions. The first one was the reconstruction loss optimising a similarity function over the intensity values of the medical volume $\mathcal{L}_{sim}$. For our experiments, we used the mean square error function and normalized cross correlation, depending on the experiment, between the warped image M^{warp} and the fixed image F. The second loss integrated anatomical knowledge by optimising the dice coefficient between the warped segmentation and the segmentation of the fixed volume: $\mathcal{L}_{sup} = Dice(M_{seg}^{warp}, F_{seg})$. Finally, a regularisation loss was also integrated to enforce smoothness of the displacement field by keeping it close to zero deformation : $\mathcal{L}_{smo} = ||\nabla \Phi_{M \to F}||$. These losses composed our final optimization strategy calculated for both $\nabla \Phi_{M \to F}$ and $\nabla \Phi_{F \to M}$

$$\mathcal{L} = (\alpha \mathcal{L}_{sim} + \beta \mathcal{L}_{sup} + \gamma \mathcal{L}_{smo})_{M \to F} + (\alpha \mathcal{L}_{sim} + \beta \mathcal{L}_{sup} + \gamma \mathcal{L}_{smo})_{F \to M}$$

where α, β and γ were weights that were manually defined. The network was optimized using Adam optimiser with a learning rate set to $1e^{-4}$.

Regarding the implementation details, for task 3, we used batch size 2 with patch size equal to $144 \times 144 \times 144$ due to memory limitations. Our normalisation strategy included the extraction of three CT windows, which all of them are used as additional channels and min-max normalisation to be in the range $(0, 1)$. For our experiments we did not use any data augmentation and we set $\alpha = 1$, $\beta = 1$ and $\gamma = 0.01$. The network was trained on 2 Nvidia Tesla V100 with 16 GB memory, for 300 epochs for $\approx$12 h. For task 4, the batch size was set to 6 with patches of size $64 \times 64 \times 64$ while data augmentation was performed by random flip, random rotation and translation. Our normalisation strategy in this case included: $\mathcal{N}(0, 1)$ normalisation, clipping values outside of the range $[-5, 5]$ and min-max normalisation to stay to the range $(0, 1)$. The weights were set to $\alpha = 1$, $\beta = 1$ and $\gamma = 0.1$ and the network was trained on 2 Nvidia GeForce GTX 1080 GPUs with 12 GB memory for 600 epochs for $\approx$20 h.

The segmentation network, used to produce noisy segmentations, was a 3D UNet trained with batch size 6, learning rate $1e^{-4}$, leaky ReLU activation functions, instance normalisation layers and random crop of patch of size $144 \times 144 \times 144$. During inference, we kept the ground truth segmentations of the organs available, we applied a normalisation with connected components and we checked each segmentations manually to remove outlier results.

3   Experimental Results

For each task, we performed an ablation study to evaluate the contribution of each component and task 3, we performed a supplementary experiment integrating the noisy labels during the pretraining. The evaluation was performed in terms of Dice score, 30% of lowest Dice score, Hausdorff distance and standard deviation of the log Jacobian. These metrics evaluated the accuracy and robustness of the method as well as the smoothness of the deformation. Our results are summarised in Table 1, while some qualitative results are represented in Fig. 2. For the inference on the test set, we used our model trained on both training and validation datasets. Concerning the computational time, our approach needs 6.21 and 1.43 s for the inference respectively for task 3 and 4. This is slower than other participants to the challenge, probably due to the size of our deep network which have around 20 millions parameters.

Concerning task 3, one can observe a significant boost on the performance when the pretraining with the noisy labels was integrated. Due to the challenging nature of this registration problem, the impact of the symmetric training was not so high in any of the metrics. On the other hand, for task 4, the symmetric component with the pretraining boosted the robustness of the method while the pretraining had a lower impact than on task 3. One possible explanation is that for this task, the number of provided volumes in combination with the nature of the problem was enough for training a learning based registration method.

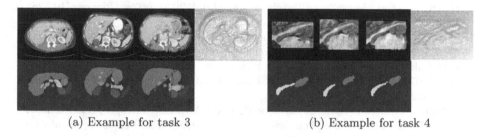

(a) Example for task 3   (b) Example for task 4

Fig. 2. Results obtained on the validation set. From left to right: moving, fixed, deformed images and the deformation grid. For the task 3, we displayed an axial view with the different organs (second row). For the task 4, we displayed a sagittal view with the head and tail masks (second row)

Table 1. Evaluation of our method for the Tasks 3 & 4 of Learn2Reg Challenge on the validation set (val) and on the test set (test).

Dataset		Task 3				Task 4			
		Dice	Dice30	Hd95	StdJ	Dice	Dice30	Hd95	StdJ
Val	Unregistered	0.23	0.01	46.1		0.55	0.36	3.91	
Val	Baseline	0.38	0.35	45.2	1.70	0.80	0.78	2.12	**0.067**
Val	Baseline + sym.	0.40	0.36	45.7	1.80	0.83	0.82	1.68	0.071
Val	Baseline + sym. + pretrain	0.52	0.50	42.3	**0.32**	**0.84**	**0.83**	**1.63**	0.093
Val	Baseline + sym. + pretrain + noisy labels	**0.62**	**0.58**	**39.3**	1.77				
Test	Baseline + sym. + pretrain + noisy labels	0.64	0.40	37.1	1.53	0.85	0.84	1.51	0.09

4 Conclusions

In this work, we summarise our method that took the 3rd place in the Learn2Reg challenge, participating on the tasks 3 & 4. Our formulation is based on spatial gradients and explores the impact of symmetry, pretraining and integration of public available datasets. In the future, we aim to further explore symmetry in our method and investigate ways that our formulation could hold diffeomorphic properties. Finally, adversarial training is also something that we want to explore in order to be deal with multimodal registration.

References

1. Avants, B.B., Epstein, C.L., Grossman, M., Gee, J.C.: Symmetric diffeomorphic image registration with cross-correlation: evaluating automated labeling of elderly and neurodegenerative brain. Med. Image Anal. **12**(1), 26–41 (2008)
2. Avants, B.B., Tustison, N., Song, G.: Advanced normalization tools (ANTS). Insight J. **2**(365), 1–35 (2009)
3. Çiçek, Ö., Abdulkadir, A., Lienkamp, S.S., Brox, T., Ronneberger, O.: 3D U-Net: learning dense volumetric segmentation from sparse annotation. In: Ourselin, S., Joskowicz, L., Sabuncu, M.R., Unal, G., Wells, W. (eds.) MICCAI 2016. LNCS, vol. 9901, pp. 424–432. Springer, Cham (2016). https://doi.org/10.1007/978-3-319-46723-8_49
4. Dalca, A., et al.: Learn2reg - the challenge, March 2020. https://doi.org/10.5281/zenodo.3715652
5. Dalca, A.V., Balakrishnan, G., Guttag, J., Sabuncu, M.R.: Unsupervised learning for fast probabilistic diffeomorphic registration. In: Frangi, A.F., Schnabel, J.A., Davatzikos, C., Alberola-López, C., Fichtinger, G. (eds.) MICCAI 2018. LNCS, vol. 11070, pp. 729–738. Springer, Cham (2018). https://doi.org/10.1007/978-3-030-00928-1_82
6. Estienne, T., Lerousseau, M., Vakalopoulou, M., Alvarez Andres, E., Battistella, E., et al.: Deep learning-based concurrent brain registration and tumor segmentation. Front. Comput. Neurosci. **14**, 17 (2020)
7. Gibson, E., Giganti, F., Hu, Y., Bonmati, E., Bandula, S., et al.: Automatic multi-organ segmentation on abdominal CT with dense V-networks. IEEE Trans. Med. Imaging **37**(8), 1822–1834 (2018)
8. Guo, Y., Wu, X., Wang, Z., Pei, X., Xu, X.G.: End-to-end unsupervised cycle-consistent fully convolutional network for 3D pelvic CT-MR deformable registration. J. Appl. Clin. Med. Phys. **21**, 193–200 (2020)
9. Haskins, G., Kruger, U., Yan, P.: Deep learning in medical image registration: a survey. Mach. Vis. Appl. **31**(1), 8 (2020). https://doi.org/10.1007/s00138-020-01060-x
10. Heinrich, M.P., Jenkinson, M., Brady, M., Schnabel, J.A.: MRF-based deformable registration and ventilation estimation of lung CT. IEEE Trans. Med. Imaging **32**(7), 1239–1248 (2013)
11. Heller, N., Sathianathen, N., Kalapara, A., Walczak, E., et al.: The KiTS19 challenge data: 300 kidney tumor cases with clinical context, CT semantic segmentations, and surgical outcomes. arXiv preprint arXiv:1904.00445 (2019)
12. Kim, B., Kim, J., Lee, J.-G., Kim, D.H., Park, S.H., Ye, J.C.: Unsupervised deformable image registration using cycle-consistent CNN. In: Shen, D., et al. (eds.) MICCAI 2019. LNCS, vol. 11769, pp. 166–174. Springer, Cham (2019). https://doi.org/10.1007/978-3-030-32226-7_19
13. Krebs, J., Delingette, H., Mailhé, B., Ayache, N., Mansi, T.: Learning a probabilistic model for diffeomorphic registration. IEEE Trans. Med. Imaging **38**(9), 2165–2176 (2019)
14. Marcus, D.S., Fotenos, A.F., Csernansky, J.G., Morris, J.C., Buckner, R.L.: Open access series of imaging studies: longitudinal MRI data in nondemented and demented older adults. J. Cogn. Neurosci. **22**(12), 2677–2684 (2010)
15. Roth, H.R., Farag, A., Turkbey, E., Lu, L., Liu, J., Summers, R.M.: Data from pancreas-CT. The cancer imaging archive (2016)

16. Simpson, A.L., Antonelli, M., Bakas, S., Bilello, M., Farahani, K., et al.: A large annotated medical image dataset for the development and evaluation of segmentation algorithms. arXiv preprint arXiv:1902.09063 (2019)
17. Sotiras, A., Davatzikos, C., Paragios, N.: Deformable medical image registration: a survey. IEEE Trans. Med. Imaging 32(7), 1153–1190 (2013)
18. Stergios, C.: Linear and deformable image registration with 3D convolutional neural networks. In: Stoyanov, D., et al. (eds.) RAMBO/BIA/TIA -2018. LNCS, vol. 11040, pp. 13–22. Springer, Cham (2018). https://doi.org/10.1007/978-3-030-00946-5_2
19. de Vos, B.D., Berendsen, F.F., Viergever, M.A., Sokooti, H., Staring, M.,Išgum, I.: A deep learning framework for unsupervised affine and deformable image registration. Med. Image Anal. 52, 128–143 (2019)

Multi-step, Learning-Based, Semi-supervised Image Registration Algorithm

Marek Wodzinski$^{(\boxtimes)}$ (ID)

Department of Measurement and Electronics,
AGH University of Science and Technology, Krakow, Poland
wodzinski@agh.edu.pl

Abstract. This paper presents a contribution to the Learn2Reg challenge organized jointly with the MICCAI 2020, more specifically, to the task related to inter-patient hippocampus registration in magnetic resonance images. The proposed algorithm is a multi-step, learning-based, and semi-supervised procedure. The method consists of a sequentially stacked U-Net-like architecture, trained in alternation. The method was ranked as the second-best (for the hippocampus registration task) in terms of the combined challenge evaluation criteria.

Keywords: Image registration · Deep learning · Medical imaging · L2R · Learn2Reg

1 Introduction

This paper presents a contribution to the Learn2Reg challenge organized jointly with the MICCAI 2020 conference [1]. The method aims to perform inter-patient registration of the hippocampus in mono-modal magnetic resonance images (MRI). The proposed method is a multi-step, learning-based algorithm. It combines the self-supervision, based on the MIND-loss [2], with the weak supervision using available segmentation masks.

2 Methods

2.1 Method

The proposed method is a learning-based, multi-step registration procedure. It consists of a sequentially stacked U-Net-like architecture [3]. The algorithm starts with initializing an identity transformation and concatenating the source and target images. The images are passed through the network and the calculated displacement field is composed with the current transformation. Then, the source image is warped with the displacement field, again concatenated with the target image, and passed to the following level (Fig. 1). The process is repeated

N. Shusharina et al. (Eds.): ABCs 2020/L2R 2020/TN-SCUI 2020, LNCS 12587, pp. 94–99, 2021.
https://doi.org/10.1007/978-3-030-71827-5_12

for a predefined number of levels. The models do not share weights and are trained simultaneously and independently with different regularization parameters. As a result, the later levels get slightly different inputs during each epoch. It was experimentally verified that this approach serves as a self-augmentation, it decreases the results for the training set but improves them for the validation set, compared to training the networks sequentially or simultaneously. The leaky ReLU was used as the activation function. The networks were trained with a small batch size, thus the batch normalization was replaced by the group normalization [4].

The objective function is a weighted sum of the modality independent neighbourhood descriptor self-similarity context (MIND-SSC) [2,5], the diffusion regularizer, and the mean squared error between the segmentation masks:

$$C(S, T, S_m, T_m, u) = MIND_{SSC}(S \circ u, T) + \alpha R(u) + \beta MSE(S_m \circ u, T_m), \quad (1)$$

where S, T, S_m, T_m denotes the source, target, source mask, and target mask respectively, u is the dense displacement field, R denotes the diffusive regularization, and α, β are the parameters controlling the transformation smoothness and the influence of segmentation masks respectively. The segmentation masks were warped with the linear interpolation to make the cost function differentiable with respect to the transformation grid.

2.2   Dataset and Experimental Setup

The dataset consists of MRI scans acquired in 195 adult subjects, both healthy and with non-affective psychotic disorders. The images show the hippocampus head and body with a small surrounding neighborhood. There are 394 volumes, 263 in the training set and 131 in the test set. The training set contains segmentation masks of the hippocampus head and body. The volumes were resampled to the same voxel resolution. A more detailed dataset description can be found in [1,6].

The training was performed for a predefined number of epochs (30), using Adam, with an exponentially decaying learning rate (initial learning rate: 0.002, decaying rate: 0.92), batch size equal to 4, and 3 network levels. A single network consists of 2 660 739 trainable parameters. The total number of trainable parameters is equal to the number of levels times the value above (7 982 217). The value for β was constant and set to 0.8, while the values for α were equal to 2.0, 2.6 and 3.4 for each level respectively. The MIND-SSC radius and dilation were set to 2. The networks were trained in alternation. This means that the objective function was evaluated, and the weights were updated independently between levels, while the other networks were in forward mode only. The motivation behind this approach was first to reduce the memory required during training, and second to provide a slight self-augmentation mechanism. Since the outputs of the first level were changing during each epoch, the inputs to the later levels were different. It was observed that this approach slightly improved the results on the validation set (Table 1). The training time was roughly 30 h using

RTX 2080 Ti. The method was implemented using PyTorch [7] and the source code is available at [8].

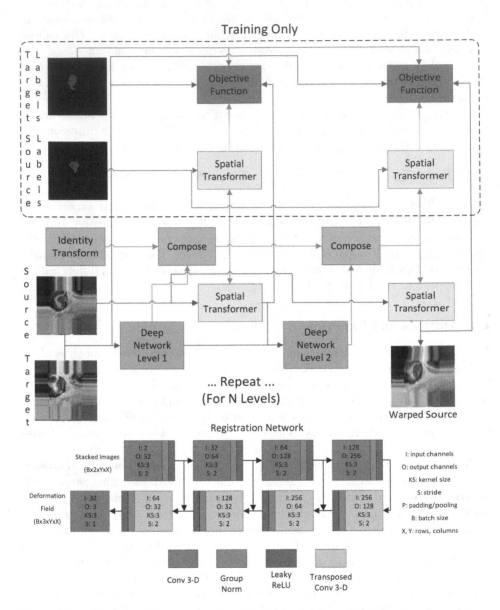

Fig. 1. Visualization of the proposed multi-step procedure and the network architecture. The segmentations masks are used during training only. Each level of the deep network consists of 5 convolutional layers (3-D) and 4 transposed convolutional layers (3-D).

3   Results

The method was evaluated using the following metrics: (i) Dice similarity coefficient of segmentation masks (DSC), (ii) robustness score (DSC30 - 30% cases with the lowest DSC), (iii) 95% percentile of Hausdorff distance of segmentations (HD95), and (iv) standard deviation of log Jacobian determinant (SDlogJ).

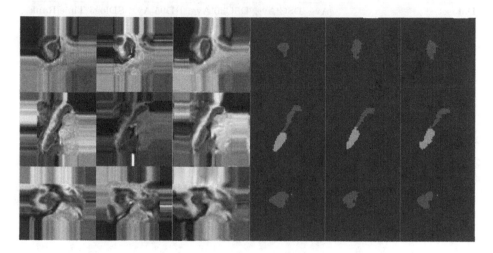

Fig. 2. Exemplary visual results of the registration for a validation pair. The columns (from left) show the middle slice of source, target, registered source, source mask, target mask, and warped source mask respectively.

The quantitative results are presented in the Table 1. The table presents also the results for the validation set with β set to 0, and the results for training the 3 levels simultaneously without alternation, and a comparison to state-of-the-art VoxelMorph [9]. An exemplary visualization of the registered images and the warped segmentation masks is shown in Fig. 2. The method was ranked as the second-best in terms of the combined challenge evaluation criteria for the hippocampus registration task [1]. A comparison to other challenge participants is presented in Table 1, is available on the challenge website [1], and will be summarized in the challenge overview article.

Table 1. Quantitative results on the validation and test sets for the proposed method, as well the other participants' methods. The high difference between the validation and test set arises from an unobserved overfitting due to a slightly incorrect experimental setup. The ranks and times are unavailable for the methods evaluated on the validation set. The "w/o alter" denotes results acquired training the 3 levels simultaneously, while the "w/o labels" shows results with β set to 0. Details about the rank calculation are available at [1].

Dataset	Avg. DSC	Avg. DSC30	Avg. HD95	Avg. SDlogJ	Time	Rank
Initial	0.55	0.36	3.91	–	–	–
VoxelMorph (val. w/o labels)	0.74	0.72	2.65	0.07	–	–
Validation (w/o alter)	0.86	0.85	1.49	0.05	–	–
Validation (w/o labels)	0.78	0.77	2.36	0.05	–	–
Validation (proposed)	0.87	0.86	1.36	0.07	–	–
Test (proposed)	0.79	0.76	2.20	0.08	0.76	2
LapIRN	0.88	0.86	1.30	0.05	1.00	1
CentraleSupelec	0.85	0.84	1.51	0.09	1.43	3
PDD-Net	0.78	0.76	2.23	0.07	0.31	4
LibReg	0.85	0.84	1.55	0.05	–	5
Deeds	0.76	0.71	2.49	0.11	3.14	6
Uppsala	0.74	0.67	2.82	0.16	21.96	7
Nifty (unoptimized)	0.73	0.65	3.37	1.00	4.72	8

4   Conclusion

The proposed method provides slightly better results than state-of-the-art unsupervised methods (Table 1). The method could be further improved by a proper augmentation of the dataset to prevent overfitting. The results for the validation set are considerably better compared to the test set because the validation pairs contained images that were used in other training pairs (a small mistake during the experiment preparation that was then repeated for other experiments to maintain consistency). It is also uncertain how the semi-supervised method will behave on a dataset acquired with different scanners or using another acquisition protocol. Noteworthy, based on the low average SDlogJ, the calculated deformations are relatively smooth and regular.

To conclude, the presented method is a semi-supervised, multi-step, learning-based registration procedure trained in alternation that outperforms state-of-the-art unsupervised methods, however, it still requires substantial improvements in terms of generalizability to new data.

Acknowledgments. This work was funded by NCN Preludium project no. UMO-2018/29/N/ST6/00143.

References

1. Dalca, A., Hering, A., Hansen, L., Heinrich, M.: The Challenge. https://learn2reg.grand-challenge.org/
2. Heinrich, M., et al.: MIND: modality independent neighbourhood descriptor for multi-modal deformable registration. Med. Image Anal. **16**(7), 1423–1435 (2012)
3. Ronneberger, O., Fischer, P., Brox, T.: U-Net: convolutional networks for biomedical image segmentation. In: Navab, N., Hornegger, J., Wells, W.M., Frangi, A.F. (eds.) MICCAI 2015. LNCS, vol. 9351, pp. 234–241. Springer, Cham (2015). https://doi.org/10.1007/978-3-319-24574-4_28
4. Wu, Y., He, K.: Group normalization. arXiv:1803.084943 (2018)
5. Heinrich, M.P., Hansen, L.: Highly accurate and memory efficient unsupervised learning-based discrete CT registration using 2.5d displacement search. In: Martel, A.L., et al. (eds.) MICCAI 2020. LNCS, vol. 12263, pp. 190–200. Springer, Cham (2020). https://doi.org/10.1007/978-3-030-59716-0_19
6. Simpson, A., et al.: A large annotated medical image dataset for the development and evaluation of segmentation algorithms. arXiv:1902.09063 (2019)
7. Paszke, A., et al.: Pytorch: an imperative style, high-performance deep learning library. In: Wallach, H., Larochelle, H., Beygelzimer, A., d'Alché-Buc, F., Fox, E., Garnett, R. (eds.) Advances in Neural Information Processing Systems, vol. 32, pp. 8024–8035. Curran Associates, Inc. (2019)
8. Wodzinski, M.: The Source Code. https://github.com/lNefarin/L2R-Task4
9. Balakrishnan, G., Zhao, A., Sabuncu, M., Guttag, J., Dalca, A.: VoxelMorph: a learning framework for deformable medical image registration. IEEE Trans. Med. Imaging **38**(8), 1788–1800 (2019)

Using Elastix to Register Inhale/Exhale Intrasubject Thorax CT: A Unsupervised Baseline to the Task 2 of the Learn2Reg Challenge

Constance Fourcade[1,2(✉)], Mathieu Rubeaux[2], and Diana Mateus[1]

[1] Ecole Centrale de Nantes, LS2N, UMR CNRS 6004, 44100 Nantes, France
{constance.fourcade,diana.mateus}@ec-nantes.fr
[2] Keosys Medical Imaging, 44300 Saint Herblain, France
mathieu.rubeaux@keosys.com

Abstract. As part of MICCAI 2020, the Learn2Reg registration challenge was proposed as a benchmark to allow registration algorithms comparison. The task 2 of this challenge consists in intrasubject 3D HRCT inhale/exhale thorax images registration. In this context, we propose a classical iterative-based registration approach based on Elastix toolbox, optimizing normalized cross-correlation metric regularized by a bending energy penalty term. This conventional registration approach, as opposed to novel deep learning techniques, reached visually interesting results, with a target registration error of 6.55 ± 2.69 mm and a Log-Jacobian standard deviation of 0.07 ± 0.03. The code is publicly available at: https://github.com/fconstance/Learn2Reg_Task2_SimpleElastix.

Keywords: Registration · Inhale/exhale lung CT

1 Introduction

Learn2Reg MICCAI 2020 satellite event is a registration challenge [3] consisting of four different tasks covering a wide range of medical image registration topics: multi-modality, noisy annotations, small datasets and large deformations. We concentrated on task 2, which consists in intrasubject inhale/exhale lung CT scans registration.

Many deep learning-based registration methods have been developed recently [1,2,9] to reduce computational time and obtain more accurate deformations. However, according to [4], conventional registration methods still reach better accuracy on inhale/exhale lung CT [8] than learning-based ones. Hence, we present results obtained using an iconic-based registration method build from Elastix toolbox [6]. Since the available training data volume is relatively small, we believe a method not requiring prior training would be efficient. In this way, we also provide to the community a well optimised baseline for comparison.

This work is partially financed through "Programme opérationnel regional FEDER-FSE Pays de la Loire 2014–2020" n°PL0015129 (EPICURE).

N. Shusharina et al. (Eds.): ABCs 2020/L2R 2020/TN-SCUI 2020, LNCS 12587, pp. 100–105, 2021.
https://doi.org/10.1007/978-3-030-71827-5_13

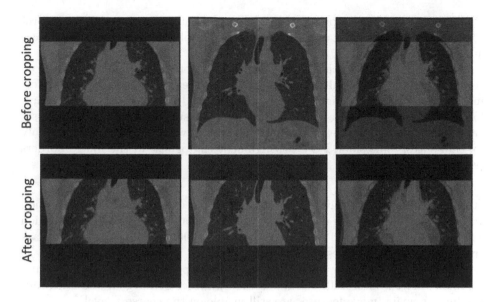

Fig. 1. Exhale, inhale and overlay of both images (1st, 2nd and 3rd column respectively) before and after (1st and 2nd row) cropping the inhale image. *Best viewed in color.*

The paper is organized as follows: we present in the next section the approach we developed as well as the dataset and the evaluation metrics provided by challenge organizers, then we expose the results, before discussing them in the conclusion section.

2   Material and Method

2.1   Dataset

The dataset of the task 2 of Learn2Reg challenge consists of 60 monocentric thorax CT images from 30 subjects [5]. A pair of images (inhale and exhale scans) is available per subject (see Fig. 1). A segmentation mask of the lungs is also available for each volume. Images size is $192 \times 192 \times 208$ and spacing is $1.75 \times 1.25 \times 1.75$ mm. Even if the dataset was split by challenge organizers into 17 training, 3 validation and 10 testing pairs, we treat every image independently, since we do not use training.

2.2   Method

The challenge objective of the second task is to register the inhale – moving – image to the exhale – fixed – one.

Preprocessing. Images provided by challenge organizers were already preprocessed to the same spatial dimension and voxel resolution. Moreover, they were affinely pre-registered.

As visible on Fig. 1, the field-of-view (FOV) of exhale images is reduced compared to the one of inhale images. To reduce registration unrealistic deformations, we decided to align the FOV of both fixed and moving images. Thus, for each exhale image slice where the body of the subject in not visible, to set values of the corresponding inhale slice to 0. Since we modify voxel values only for null slices, some small FOV misalignment can persist in the image borders (see 3rd column, Fig. 2).

Registration. Our iterative registration method uses a B-spline transformation with four resolutions, each dividing the previous image size by two. Registration resolutions were optimized minimizing Eq. 1 for 1000 iterations, using an adaptive stochastic gradient descent optimizer [6].

These hyperparameter choices did not need special tuning and are quite common. Indeed, after 1000 iterations the optimizer converged, and with four resolutions, main image features are still visible on the coarsest level.

$$C(\mathbf{T}) = -C_{similarity}(I_{fixed}, \mathbf{T}(I_{moving})) + \lambda C_{smooth}(\mathbf{T}) \qquad (1)$$

Since both fixed and moving images, I_{fixed} and I_{moving}, are from the same modality, $C_{similarity}$ corresponds to the normalized cross-correlation metric. Moreover, to ensure smooth and realistic-looking deformations, the similarity metric is regularized by C_{smooth}, a bending energy penalty term [7]. It corresponds to the second derivative of the transformation $\mathbf{T}$. In our experiments, several values of λ were tested (0, 0.1, 1, 10). We choose $\lambda = 1$, to balance registration accuracy and smoothness.

2.3   Evaluation Metrics

Two evaluation metrics were used for the challenge. The target registration error (**TRE**) evaluates registration precision, while the standard deviation of the logarithm of the Jacobian of the deformation field (**SDLogJ**) quantifies registration smoothness. The TRE is computed on 100 landmarks, automatically set on fixed images, while correspondences in moving images are manually annotated. The landmarks are on lungs salient features, like vessels or nodules. For both metrics, a lower value indicates a better registration.

3   Results

With the proposed Elastix method, we obtain a TRE of 6.55 ± 2.69 mm and a SDLogJ of 0.07 ± 0.03 on the test dataset. More details subject-wise are visible Table 1. For reference, the initial error between fixed and moving images (after

Table 1. Results on the 10 subject from the test dataset

Subject	1	2	3	4	5	6	7	8	9	10	Mean
TRE	7.78	6.84	7.42	5.19	9.57	11.70	5.19	1.35	5.61	4.82	6.55 ± 2.69
SDLogJ	0.08	0.17	0.07	0.06	0.07	0.07	0.04	0.05	0.06	0.07	0.07 ± 0.03

affine registration only) was 10.24 ± 2.72 mm. Table 2 shows these results, along with the ones of the participating deep learning-based methods.

Our low value of SDlogJ confirms that the proposed method creates smooth deformations and realistic-looking images. However, the high TRE value reflects a lack of precise registration inside the lungs, as illustrated 6th row of Fig. 2.

Table 2. Comparison of our results with the ones obtained using deep learning-based approaches for the task 2 of the Learn2Reg challenge

	TRE	SDLogJ
Identity	10.24	0.00
Tony (LapIRN)	3.24	0.06
Lasse (PDD-Net)	2.46	0.07
Niklas (Uppsala)	9.00	0.12
Elastix (ours)	**6.55**	**0.07**

4   Discussion and Conclusion

In this paper, we propose a conventional registration method for the Learn2Reg challenge task 2, based on a normalized cross-correlation and a bending energy regularization term.

Both fixed and moving images of the dataset of the task 2 did not present the same FOV. Hence, we cropped images slices to obtain more realistic registration results. Without performing this prior FOV alignment, our method performs poorly, as illustrated on image borders by red boxes, last row of Fig. 2. Even if our approach should be adapted from a dataset to another, it seems important to align the FOV of fixed and moving images to reach better registration results.

With regards to the learning based methods, our computation time is longer, yet we do not require time-expensive prior training. Also, our results reach similar smoothness, but slightly lower accuracy over the testing dataset. Regarding the use of hyperparameters, both conventional and learning-based methods should find appropriate values for the number of resolutions/network depth, the number of iterations/number of epochs, the regularization weight, etc.

Overall, our approach provides reasonable registration results, although inside the lungs the TRE remains high. This could be improved in a second

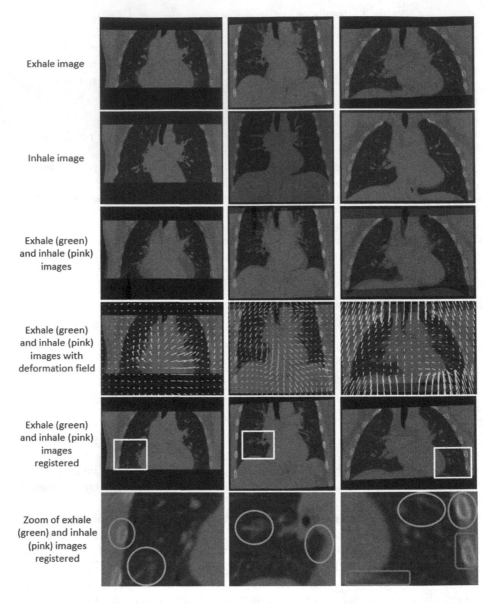

Fig. 2. For three validation subjects (resp. each column), exhale image, inhale image, overlay of exhale and inhale image, overlay of exhale and inhale images with the deformation field from inhale to inhale in green and overlay of exhale and registered inhale images (1ˢᵗ, 2ⁿᵈ, 3ʳᵈ, 4ᵗʰ and 5ᵗʰ row resp.). The 6ᵗʰ row zooms inside white squares of the 5ᵗʰ row images. In this last row, body parts visually seem accurately registered (blue circles – gray-scale colors), but registration approximations are visible within the inner lung regions (orange circles – pink-green colors). Larger registration errors due to misaligned initial fields of view are highlighted in the bottom right image by red boxes. *Best viewed in color.* (Color figure online)

step, masking the body and focusing the registration only on the lungs. We provide all the implementation details, hyperparameters, and code, such that the proposed method could serve as a non-learning based baseline for comparison.

References

1. Balakrishnan, G., Zhao, A., Sabuncu, M.R., Guttag, J., Dalca, A.V.: VoxelMorph: a learning framework for deformable medical image registration. IEEE Trans. Med. Imaging **38**(8), 1788–1800 (2019)
2. Casamitjana, A., et al.: Introduction to medical image registration with DeepReg, between old and new. arXiv preprint arXiv:2009.01924 (2020)
3. Dalca, A., Hering, A., Hansen, L., Heinrich, M.: Learn2Reg Challenge (2020). https://learn2reg.grand-challenge.org/
4. Heinrich, M.P., Hansen, L.: Highly accurate and memory efficient unsupervised learning-based discrete CT registration using 2.5D displacement search. In: Martel, A.L., et al. (eds.) MICCAI 2020. LNCS, vol. 12263, pp. 190–200. Springer, Cham (2020). https://doi.org/10.1007/978-3-030-59716-0_19
5. Hering, A., Murphy, K., van Ginneken, B.: Lean2Reg Challenge: CT Lung Registration - Training Data (2020). https://zenodo.org/record/3835682#.X2Nklos69hE
6. Klein, S., Staring, M., Murphy, K., Viergever, M., Pluim, J.: elastix: a toolbox for intensity-based medical image registration. IEEE Trans. Med. Imaging **29**(1), 196–205 (2010)
7. Rueckert, D., Sonoda, L.I., Hayes, C., Hill, D.L.G., Leach, M.O., Hawkes, D.J.: Non-rigid registration using free-form deformations: application to breast MR images. IEEE Trans. Med. Imaging **18**(8), 712–721 (1999)
8. Ruhaak, J., et al.: Estimation of large motion in lung CT by integrating regularized keypoint correspondences into dense deformable registration. IEEE Trans. Med. Imaging **36**(8), 1746–1757 (2017)
9. de Vos, B.D., Berendsen, F.F., Viergever, M.A., Sokooti, H., Staring, M., Išgum, I.: A deep learning framework for unsupervised affine and deformable image registration. Med. Image Anal. **52**, 128–143 (2019)

TN-SCUI – Thyroid Nodule Segmentation and Classification in Ultrasound Images

Automatic Segmentation and Classification of Thyroid Nodules in Ultrasound Images with Convolutional Neural Networks

Mingyu Wang, Chenglang Yuan, Dasheng Wu, Yinghou Zeng, Shaonan Zhong, and Weibao Qiu$^{(\boxtimes)}$

Institute of Biomedical and Health Engineering, Shenzhen Institutes of Advanced Technology, Chinese Academy of Sciences, Beijing, China
wb.qiu@siat.ac.cn

Abstract. Ultrasound image plays an important role in the diagnosis of thyroid disease. Accurate segmentation and classification of thyroid nodules are challenging due to their heterogeneous appearance. In this paper, we propose an efficient cascaded segmentation framework and a dual-attention ResNet-based classification network to automatically achieve the accurate segmentation and classification of thyroid nodules, respectively. We evaluate our methods on the training dataset TN-SCUI 2020 Challenge. The 5-fold cross validation results demonstrate that the proposed methods achieve average IoU of 81.43% in segmentation task, and average F1 score of 83.22% in classification task. Finally, our method ranks the first place of segmentation task on the test set through the final online verification. The source code of the proposed methods is available at https://github.com/WAM AWAMA/TNSCUI2020-Seg-Rank1st.

Keywords: Thyroid nodule · Ultrasound image · Convolutional neural network

1 Introduction

Thyroid nodules are one of most commonly diagnosed nodular lesions in the adult population. The nodules, detected at an early stage, are an extremely curable disease, and thus an accurate differentiation between malignant and benign thyroid nodules is necessary to ensure proper clinical management of malignant nodules [1]. Because the ultrasound imaging is noninvasive, realtime, and radiation-free, it is the key tool for diagnosis of thyroid nodules. However, the cumulative errors collected from blurring boundaries and significant changes in the appearance or intensity of thyroid nodules among different ultrasound images, make it challenging to analyze and recognize the subtle difference between malignant and benign nodules [2]. In this paper, we propose an

---

M. Wang and C. Yuan—Contributed equally to this work.

Electronic supplementary material The online version of this chapter (https://doi.org/10.1007/978-3-030-71827-5_14) contains supplementary material, which is available to authorized users.

© Springer Nature Switzerland AG 2021
N. Shusharina et al. (Eds.): ABCs 2020/L2R 2020/TN-SCUI 2020, LNCS 12587, pp. 109–115, 2021.
https://doi.org/10.1007/978-3-030-71827-5_14

efficient cascaded segmentation network and a dual-attention ResNet-based classification network to achieve automatic and accurate segmentation and classification of thyroid nodules, respectively.

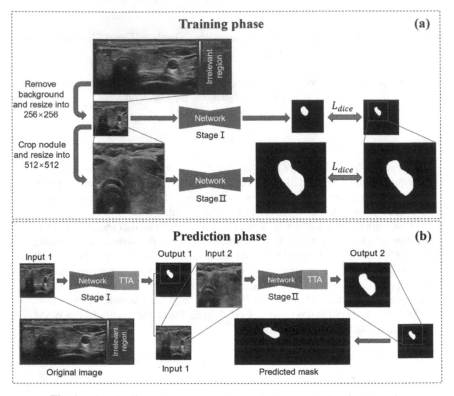

Fig. 1. The pipeline of our proposed cascaded segmentation framework.

2   Method

2.1   Data Preprocessing

Due to different acquisition protocols, some thyroid ultrasound images have irrelevant regions (as shown in Fig. 1). First, we remove these regions which may bring redundant features by using a threshold-based approach. Specifically, we perform the operation of averaging along the x and y axes on original images with pixel values from 0 to 255, respectively, after which rows and columns with mean values less than 5 are removed. Then the processed images are resized to 256×256 pixels as the input of the first segmentation network.

2.2 Cascaded Segmentation Framework

Our cascaded segmentation pipeline is shown in Fig. 1. We train two networks which share the same encoder-decoder structure with Dice loss function. The first segmentation network (at stage I of cascade) is trained to provide the rough localization of nodules, and the second segmentation network (at stage II of cascade) is trained for fine segmentation based on the rough localization. To our knowledge, in some current cascaded segmentation frameworks, the real output (mask or probability map) or pseudo-label output of the first network is generally fed for training the second network, so that the second network gets contextual information [3, 4]. But our preliminary experiments show that the provided context information in first network may do not play a significant auxiliary role for refinement of the second network. Therefore, we only train the second network using images within region of interest (ROI) obtained from ground truth (GT).

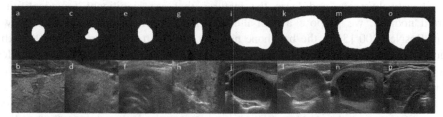

Fig. 2. Overview of thyroid nodules in the training dataset of TN-SCUI. a to h, small nodules (in the case of picture size 256 × 256, the minimum external square edge length of the nodule is less than 80 pixels), i to p, large nodules (edge length is greater than 80 pixels). The upper row: GT. The lower row: ultrasound image corresponding to GT.

When training the second network, we expand the nodule ROI obtained from GT, then the image in the expanded ROI is cropped out and resized to 512 × 512 pixels for training the second network. We observe that, in most cases, the large nodules generally have clear boundaries, and the gray value of small nodules is quite different from that of surrounding normal thyroid tissue (Fig. 2). Therefore, background information (the tissue around the nodule) is significant for segmenting small nodules. As shown in Fig. 3, in the preprocessed image with the size of 256 × 256 pixels, the minimum external square of the nodule ROI is obtained first, and then the external expansion m is set to 20 if the edge length n of the square is greater than 80, otherwise the m is set to 30.

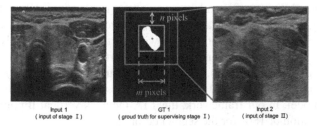

Fig. 3. Example of input image of second segmentation network. n, edge length of minimum external square of the nodule ROI. m, number of expanded pixels.

2.3   Dual-Attention ResNet Framework

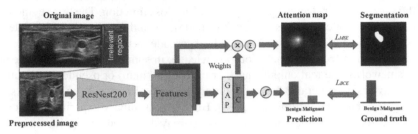

Fig. 4. The pipeline of our proposed Dual-attention ResNet framework.

To exhaustively focus and learn the features that are significant for identifying thyroid nodules, we propose a dual-attention ResNet framework. Specifically, we adopt ResNeSt200 [5] as the backbone network architecture to perform the classification of thyroid nodules. ResNeSt introduces the Split-Attention block, which enables feature map attention across different feature map groups, and spontaneously improves the learned feature representations to boost model performance.

In addition, we employ an online attentional mechanism based on class activation mapping to focus on the intrinsic relationship between the feature information of thyroid nodules and their clinical characteristics. The core idea is to convey weights of the fully connected layer onto the convolutional feature maps for generating the attention maps and optimizing dynamically network's decision. As shown in Fig. 4, we define F as the feature maps before the global average pooling operation and W as the weight matrix of the fully connected layer. The generating attention maps is defined as: $AM = Norm\left(\sum_{c=1} (F * W)\right)$, where the Norm denotes the normalization process and the operation of summation is performed along the channel axis. Finally, our loss function consists of two parts. On the one hand, we use MSE loss between attention maps and ground truths of segmentation to constrain the model's attention to the location of the thyroid nodules. On the other hand, we choose BCE loss between predictions and ground truths of classification to reduce confidence errors.

2.4   Data Augmentation and Test Time Augmentation

In both two tasks, following methods are performed in data augmentation: 1) horizontal flipping, 2) vertical flipping, 3) random cropping, 4) random affine transformation, 5) random scaling, 6) random translation, 7) random rotation, and 8) random shearing transformation. In addition, one of the following methods was randomly selected for additional augmentation: 1) sharpening, 2) local distortion, 3) adjustment of contrast, 4) blurring (Gaussian, mean, median), 5) addition of Gaussian noise, and 6) erasing.

Test time augmentation (TTA) generally improves the generalization ability of the segmentation model. In our framework, the TTA includes vertical flipping, horizontal flipping, and rotation of 180° for the segmentation task.

3   Experiments

We validate our method on the training dataset of TN-SCUI 2020 challenge, which includes 2003 malignant nodules and 1641 benign nodules.

Cross Validation with a Size and Category Balance Strategy. 5-fold cross validation was used to evaluate the performance of our proposed method. In our opinion, it is necessary to keep the size and category distribution of nodules similar in the training and validation sets. In practice, the size of a nodule is the number of pixels of the nodule after unifying preprocessed image to 256×256 pixels. We stratified the size into three grades: 1) less than 1722 pixels, 2) less than 5666 pixels and greater than 1722 pixels, and 3) greater than 5666 pixels. These two thresholds, 1722 pixels and 5666 pixels, were close to the tertiles, and the size stratification was statistically significantly associated with the benign and malignant categories by the Chi-square test ($p < 0.01$). We divided images in each size grade group into 5 folds and combined different grades of single fold into new single fold. This strategy ensured that final 5 folds had similar size and category distributions.

Implementation Details. We implemented our framework in PyTorch using 3 NVIDIA GTX 1080TI GPUs. For both two tasks, we chose Adam as the optimizer, and used the learning rate strategy based on learning rate warm-up with 5 epochs and cosine decay with 350 epochs. In the segmentation task, all network frameworks were built by segmentation_models_pytorch[1], the batch sizes were 10 and 3 for the first and second network, respectively. Learning rate increased from $1e-12$ to $1e-4$ during warm-up phase, and then gradually decreased to $1e-12$ during cosine-decay phase. When training the segmented network with Efficientnet as encoder, we initialized the encoder using weights pretrained on ImageNet. In the classification task, the batch size was 16, and the learning rate increased from $1e-12$ to $2.5e-3$ during warm-up phase, and then gradually decreased to $1e-12$ during cosine-decay phase. We employed Dice similarity coefficient (DSC) and intersection over union (IoU) to evaluate the segmentation model performance. And area under the receiver operating characteristic curve (AUROC), sensitivity (SEN), specificity (SPE), accuracy (ACC) and F1 score were calculated to evaluate the classification model performance. Several experiments are implemented for extensive comparisons.

4   Results

4.1   Segmentation Results

We tested different network structures on the validation set in the first fold, and the segmentation results without TTA are shown in Table 1. In order to make the test results as close to the real situation as possible (such as testing on the final test set of TN-SCUI challenge), all the indicators in the table were calculated based on the GT corresponding to the original image, so the predicted masks were restored to the size of the unprocessed

---

[1] https://github.com/qubvel/segmentation_models.pytorch.

original images. The results show that among different networks, DeeplabV3plus with pretrained Efficientnet-B6 encoder work best, reaching 0.8699 for DSC and 79.00% for IoU.

Table 1. Segmentation results of different networks on the validation set in the first fold.

Network	Weight initialization	DSC	IOU (%)
U-net	Random	0.8493	76.53
U-net++	Random	0.8506	76.82
U-net (ef-b6[a])	ImageNet[b]	0.8537	77.58
U-net (ef-b6[a])+scSE[c]	ImageNet[b]	0.8609	78.33
Deeplabv3Plus (ef-b6[a])	ImageNet[b]	**0.8699**	**79.00**

a, network with Efficientnet-B6 as encoder; b, encoder was initialized by the weights pretrained on ImageNet; c, spatial and channel-wise squeeze & excitation attention module.

We choose DeeplabV3plus with pretrained Efficientnet-b6 encoder as the two segmentation networks for cascading, and the cascaded segmentation results are shown in Table 2. All metrics in the table were calculated using the GT corresponding to the original image. The results of the first fold show that by using cascades and simultaneously using TTA in both first and second networks, IoU improved from 79.00% to 81.44% and DSC improved from 0.8699 to 0.8864 continuously. The results of the remaining 4 folds also show the same trend (as shown in Supplementary Table 1). Finally, we get an average DSC of 0.8873 and average IoU of 81.43% in 5-fold cross-validation.

Table 2. Ablations study of proposed cascaded segmentation framework on the validation set in the first fold.

Stage I	TTA at stage I	Stage II	TTA at stage II	DSC	IoU (%)
✓				0.8699	79.00
✓	✓			0.8775	80.01
✓		✓		0.8814	80.75
✓	✓	✓		0.8841	81.05
✓		✓	✓	0.8840	81.16
✓	✓	✓	✓	**0.8864**	**81.44**

4.2 Classification Results

As shown in Table 3, due to the powerful feature representation ability, ResNest200 achieve 0.8186 F1 score, exhibiting more excellent performances than ResNet and Efficientnet. Moreover, the online attention mechanism bring another 0.0136 improvement in F1 score because of the effective guidance towards important activation positions.

Table 3. Comparisons between the proposed method and other methods in classification of thyroid nodules.

Method	AUROC	SEN	SPE	ACC	F1-score
ResNet101	0.8057	0.8105	**0.7091**	0.7647	0.7908
Efficientnet-b5	0.8260	0.8728	0.6515	0.7729	0.8083
ResNeSt50	0.8386	0.8454	0.7000	0.7798	0.8081
ResNeSt101	0.8444	0.8953	0.6303	0.7756	0.8141
ResNeSt200	0.8483	0.8778	0.6758	0.7866	0.8186
Ours	**0.8549**	**0.9027**	0.6758	**0.8003**	**0.8322**

5 Conclusions

We proposed two efficient and automatic frameworks with convolutional neural networks to segment and classify the thyroid nodules, respectively. The experimental results demonstrated that both cascade strategy and TTA could effectively improve the performance of segmentation. For the classification task, our proposed dual-attention ResNet achieved better performance than the original ResNet as well as ResNeSt.

References

1. Kitahara, C.M., Sosa, J.A.: The changing incidence of thyroid cancer. Nat. Rev. Endocrinol. **12**(11), 646–653 (2016)
2. Ying, X., et al.: Thyroid nodule segmentation in ultrasound images based on cascaded convolutional neural network. In: Cheng, L., Leung, A.C.S., Ozawa, S. (eds.) ICONIP 2018. LNCS, vol. 11306, pp. 373–384. Springer, Cham (2018). https://doi.org/10.1007/978-3-030-04224-0_32
3. Isensee, F., Maier-Hein, K.H.: An attempt at beating the 3D U-Net. arXiv preprint arXiv:1908. 02182 (2019)
4. Cheng, H.K., et al.: CascadePSP: toward class-agnostic and very high-resolution segmentation via global and local refinement. In: Proceedings of the IEEE/CVF Conference on Computer Vision and Pattern Recognition (2020)
5. Zhang, H., et al.: ResNeSt: split-attention networks. arXiv preprint arXiv:2004.08955 (2020)

LRTHR-Net: A Low-Resolution-to-High-Resolution Framework to Iteratively Refine the Segmentation of Thyroid Nodule in Ultrasound Images

Huai Chen[1], Shaoli Song[2], Xiuying Wang[3], Renzhen Wang[4], Deyu Meng[4], and Lisheng Wang[1(✉)]

[1] Institute of Image Processing and Pattern Recognition, Department of Automation, Shanghai Jiao Tong University, Shanghai 200240, People's Republic of China
lswang@sjtu.edu.cn
[2] Department of Nuclear Medicine, Fudan University Shanghai Cancer Center, Shanghai, China
[3] School of Computer Science, The University of Sydney, Sydney, Australia
[4] School of Mathematics and Statistics, Xi'an Jiaotong University, Xi'an, China

Abstract. The thyroid nodule is quickly increasing worldwide and the thyroid ultrasound is the key tool for the diagnosis of it. For the subtle difference between malignant and benign nodules, segmenting lesions is the crucial preliminary step for diagnosis. In this paper, we propose a low-resolution-to-high-resolution segmentation framework for TN-SCUI2020 challenge to alleviate the workload of clinicians and improve the efficiency of diagnosis. Specifically speaking, in order to integrate multi-scale information, several low-resolution segmenting results are obtained firstly and combined with a high-resolution image to refine them and obtain high-resolution results. Secondly, iterative-transfer is proposed to effectively initialize network based on previous trained one on small-scale images. Finally, ensemble refinement is introduced to utilize multiple models to refine the segmentation again. Experimental results showed the effectiveness of the proposed framework. And we won the 2nd place in the segmentation task of TN-SCUI2020.

Keywords: Thyroid nodule segmentation · Thyroid ultrasound · TN-SCUI2020 · Low-resolution-to-high-resolution · Iterative refinement

1 Introduction

The thyroid gland is a butterfly-shaped endocrine gland that is normally located in the lower front of the neck. It secretes indispensable hormones that are necessary for all the cells in your body to work normally [1]. The term thyroid nodule refers to an abnormal growth of thyroid cells that forms a lump within the thyroid gland [2].

© Springer Nature Switzerland AG 2021
N. Shusharina et al. (Eds.): ABCs 2020/L2R 2020/TN-SCUI 2020, LNCS 12587, pp. 116–121, 2021.
https://doi.org/10.1007/978-3-030-71827-5_15

Statistical studies show that the incidence of this disease increases with age, extending to more than 50% of the world's population. Until recently, thyroid cancer was the most quickly increasing cancer in the United States. It is the most common cancer in women 20 to 34 [3]. Although the vast majority of thyroid nodules are benign (noncancerous), a small proportion of thyroid nodules contains thyroid cancer. In order to diagnose and treat thyroid cancer at the earliest stage, it is desired to characterize the nodule accurately.

Thyroid ultrasound is a key tool for thyroid nodule evaluation. It is non-invasive, real-time and radiation-free. However, it is difficult to interpret ultrasound images and recognize the subtle difference between malignant and benign nodules. The diagnosis process is thus time-consuming and heavily depends on the knowledge and experience of clinicians [4].

The challenge of Thyroid Nodule Segmentation and Classification in Ultrasound Images (TN-SCUI 2020) [4] aims to provide a benchmark to validate all of the state-of-the-art computer-aided diagnosis (CAD) systems for thyroid nodule diagnosis.

In this paper, based on the large public dataset provided by TN-SCUI2020 [4], we propose a framework to complete the segmenting of thyroid ultrasound to assist the doctor in diagnosis. Firstly, due to the great change of lesions in size, we propose a small-resolution-to-high-resolution framework, where several low-resolution segmentation networks are firstly trained to gain results of ultrasound images under low scales. Then, these predictions will be concatenated with high-resolution images to make final results. Secondly, to reuse the power of pre-trained low-resolution networks, we propose iterative-transfer to iteratively transfer them to high-resolution ones. Finally, we propose ensemble refinement to combine outputs and high-level features of multiple models to further refine results again.

2 Methods

As illustrated in Fig. 1, our framework can be divided into two parts. The first part aims to refine high-resolution segmentation by combining multi-scale results. The second part is to effectively refine final results based on multiple models.

2.1 Refining High-Resolution Segmentations Based on Multi-scale Results

The main reasons for us to firstly get low-resolution segmentations and combine them to get final high-resolution predictions are shown as followed:

1) Thyroid nodules' lesions in ultrasound varies greatly in size, it is not suitable to set network with fixed receptive field to cope all image instances. Thus, preparing several low-resolution segmentations by low-scale images for final high-resolution results is an expected strategy. By which, network with same receptive field in each step can perform well for targets with varied size. And the final high-resolution results can refer to previous low-resolution ones.

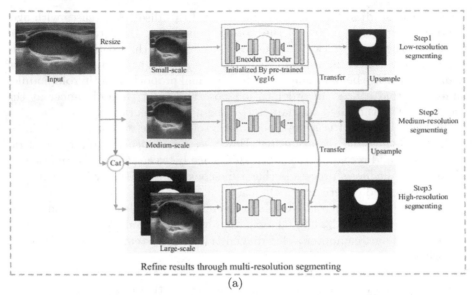

(a)

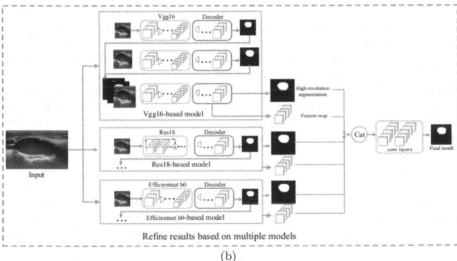

(b)

Fig. 1. An illustration of the framework. Figure (a) shows the single model, where several low-resolution segmentations will be obtained firstly and promote the final high-resolution segmenting. The iterative-transfer is utilized to initial networks based on pre-trained low-resolution networks. Figure (b) shows the ensemble refinement strategy.

2) Transferring pre-trained networks, especially models based on ImageNet, is a common trick to improve new tasks. However, when transferring models to ultrasound images, two channels of input will be redundant due to the fact that ultrasound images are gray ones. Therefore, combining two previous

results (low-resolution segmentation and medium-resolution ones) with high-scale image as the input is a good strategy to release model's performance.

It worth noting that we train networks for varied resolutions one by one (low-resolution to medium-resolution to high-resolution). And the pre-trained models for lower resolution will be the initialization for high-resolution ones, which named as iterative-transfer.

2.2 Refining Results Based on Multiple Models

Multi-model ensemble is a common strategy to unite outputs of multiple models to get final results. To better utilize models, [5] propose parallel training to fuse the medium features to make final predictions. In this paper, referring to this idea, we combine models' outputs with medium features and build a new simple network, constructed by several convolution layers, to refine predictions again.

As shown in Fig. 1(b), we build three models (based on pre-trained VGG16 [7], Res18 [6], Efficientnet-b0 [8]) and the fusion net is constructed by three '3×3' convolution layers.

3 Experiments

3.1 Dataset

TN-SCUI2020 provides 3644 training data and 910 test data, each of which is labeled by experienced doctors with pixel-level and image-level labels. We focus on the segmentation task, and we divide 360 images as validate data from the training data during the training stage.

3.2 Evaluation Metrics

Segmentation IoU score [4]: IoU score is calculated by the area of the intersection of two regions divided by the area of their union set. It is a good indicator of whether the prediction is consistent with the label.

$$IoU(Y, \bar{Y}) = \frac{TP}{TP + FN + FP} \tag{1}$$

Where Y is the ground truth and $\bar{Y}$ is the prediction. And TP, FN, FP are true positives, false negatives and false positives respectively. Considering the evaluation metric, we define the loss function as followed:

$$loss = \frac{Y \times \bar{Y}}{Y \times \bar{Y} + Y \times (1 - \bar{Y}) + (1 - Y) \times \bar{Y}} \tag{2}$$

3.3 Implementation Details

Low-Resolution Segmenting: The pre-trained model in ImageNet is utilized as the initial encoder. Under the strategy of fine-tune, the encoder is firstly frozen and only decoder is trained in the first 40 epochs and then both encoder and decoder are jointly trained in the following 100 epochs. Optimizers on these two stages are both Adam. And the learning rate is set as 10^{-3} for frozen stage and 10^{-4} for fine-tune stage. The batch size is 16.

Medium-Resolution Segmenting: The training epoch is set as 40 and the total model is initialized by the pre-trained model in Step1. The optimizer is Adam with learning rate of 10^{-4}. The batch size is 3 due to the limitation of GPU memory.

High-Resolution Segmenting: The training details are similar to Step2. However, the batch size is set as 1 due to the limitation of memory.

Others: For the training of models, the soft IoU is set as the loss function and the training rate is reduced to half if the loss does not decrease in five consecutive epochs.

4 Results

Table 1. Comparison of segmentation results on validate data.

Basebone	Initialization	Output scale	Mean Iou(%)
VGG16 (step1)	Trained on ImageNet	112×112	77.2
VGG16 (step2)	Iterative − transfer	224×224	78.4
VGG16 (step3)	Iterative − transfer	448×448	79.4
Efficientnet-b0 (step1)	trained on ImageNet	112×112	75.8
Efficientnet-b0 (step2)	Iterative − transfer	224×224	79.3
Efficientnet-b0 (step3)	Iterative − transfer	448×448	79.9
Res18 (step1)	Trained on ImageNet	112×112	77.1
Res18 (step2)	Iterative − transfer	224×224	78.6
Res18 (step3)	Iterative − transfer	448×448	79.7
Ensemble refinement	−	448×448	80.4

The segmentation results on validate data are shown on Table 1. It worth noting that the input of single model with output 448×448 is the combination of original image and two previous outputs with varied resolutions. We can come to conclusions as followed:

1) Based on the proposed iterative-transfer and low-resolution-to-high-resolution strategy, the segmentation can be improved whatever base bone is.
2) Ensemble refinement, utilizing multiple features and predictions from models, can further improve the performance.

5   Conclusion

In this paper, we propose a novel framework for the segmenting of thyroid nodule in ultrasound image. The proposed low-resolution-to-high-resolution is an effective strategy to refine the predicted results by merging low-resolution segmentations with high-scale image to predict high-resolution output. The proposed iterative-transfer is a good initialization strategy, reusing the power model trained on low-resolution images. Additionally, the ensemble refinement is a convenient mechanism to fuse features and outputs of multiple models to further refine the segmentation.

References

1. https://www.btf-thyroid.org/what-is-thyroid-disorder
2. http://www.thyroid.org/wp-content/uploads/patients/brochures/Nodules_brochure.pdf
3. https://www.cancer.net/cancer-types/thyroid-cancer/statistics
4. https://tn-scui2020.grand-challenge.org/Home/
5. Chen, H., Wang, X., Huang, Y., Wu, X., Yu, Y., Wang, L.: Harnessing 2D networks and 3D features for automated pancreas segmentation from volumetric CT images. In: Shen, D., et al. (eds.) MICCAI 2019. LNCS, vol. 11769, pp. 339–347. Springer, Cham (2019). https://doi.org/10.1007/978-3-030-32226-7_38
6. He, K., Zhang, X., Ren, S., Sun, J.: Deep residual learning for image recognition, pp. 770–778 (2016)
7. Simonyan, K., Zisserman, A.: Very deep convolutional networks for large-scale image recognition (2014)
8. Tan, M., Le, Q.V.: EfficientNet: rethinking model scaling for convolutional neural networks, pp. 6105–6114 (2019)

Coarse to Fine Ensemble Network for Thyroid Nodule Segmentation

Zhe Tang[✉] and Jianqiang Ma[✉]

Alibaba, Damo Academy, Hangzhou, China
tangzhe0220@icloud.com, mjq93@qq.com

Abstract. A thyroid nodule is defined as a small lump of tissue (either solid or cystic - filled with fluid), usually more than one-quarter of an inch in diameter, that may protrude from the neck's surface or may form in the thyroid gland itself. The nodule can be either benign (non-cancerous) or malignant (cancerous). By the age of 45, up to half of the normal people have thyroid nodules that can be seen on an ultrasound. Fortunately, about 95% of thyroid nodules are benign. Recently, artificial intelligence becomes more and more popular in medical image processing. It helps the doctor in some scenarios; one is the thyroid nodule analysis in the ultrasound image. In this competition, we design a robust coarse to fine network that reaches a high-performance nodule segmentation result. Our method could handle some problematic cases, especially with no clear boundary of the thyroid nodule.

Keywords: Thyroid nodule · Image segmentation

1 Introduction

Image segmentation is a fundamental technique in medical image processing. It is widely used to segment targets to assist doctors in making clinical decisions. For the thyroid nodule, doctors usually need to diagnosis according to the imaging manifestations features, such as the nodule location, volume size, edge smoothness, etc. Artificial intelligence could help doctors with these diagnoses so that to make a better decision for BI-RADS validation.

The quality of data strongly influences ultrasound image segmentation. Artifacts such as speckle, shadows, missing boundary, and noise usually exist, leading to nodule segmentation difficulty. In this competition, we present a novel method "Coarse to fine segmentation network", which reaches a 0.8194 IOU performance at the final leaderboard 3rd.

The rest of this paper is organized as follows. Section 'Method' describes the idea of this innovation, the architecture of the proposed network, and usage of the dataset. Section 'Experiments and Results' describes the dataset and evaluation results, section 'Conclusion' summarizes this proposed method.

N. Shusharina et al. (Eds.): ABCs 2020/L2R 2020/TN-SCUI 2020, LNCS 12587, pp. 122–128, 2021.
https://doi.org/10.1007/978-3-030-71827-5_16

2   Method

In our method, we present an ensemble image segmentation network. There are four unique algorithms. Two basic segmentation network including deeplabv3+ [1] and U2Net [3] shown as Seg. Network A and Seg. Network B in Fig. 1, while network A based on deeplabv3+ uses two backbones, xception and resnet101 for the ensemble. Another two networks C and D, are described in section 'Architecture of Coarse to Fine Network'. A two stages of image segmentation networks are used to produce coarse segmentation and fine segmentation. Feature extraction backbone is based on xception in network C and D. The basic workflow is shown in Fig. 1.

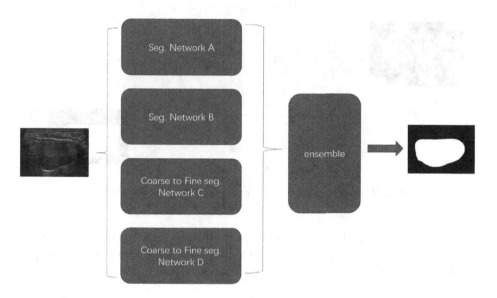

Fig. 1. Overview of work flow

2.1   Architecture of Coarse to Fine Network

Due to the difficulty of ultrasound image segmentation, especially the around nodule boundary, only an individual segmentation network can not always get a highly accurate result. Thus, we design a coarse to fine structure, which is a two-stage segmentation algorithm. The coarse to fine network consists of two main parts, coarse segmentation network, and fine segmentation network. The coarse and fine segmentation networks are based on Deeplabv3+ architecture. The coarse network outputs an initial predict mask, which is concatenated with the original image and set as the fine network's input. These two networks are optimized individually.

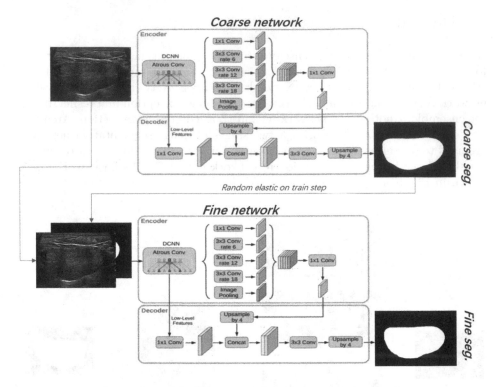

Fig. 2. Overview of coarse to fine segmentation network

As shown in Fig. 2, there are two Deeplabv3+ networks. In the first network, the origin ultrasound image with data augmentation is set as input, and output a segment predict mask, then we concatenate with image and set as the input of the fine network.

We use five cross-validations on the first coarse segmentation network, inference segment mask of each case, some of them may have good results, but some are bad. We compare with ground truth. A threshold of 0.75 was used to compare with IOU between predict coarse mask and ground truth; when IOU $\geqslant 0.75$, we set these masks as the input of fine segmentation network, while the masks with IOU < 0.75 is replaced by ground truth with applying the random elastic transformation.

When training on a fine segmentation network, an original image and a coarse mask are concatenated as double-channel input, producing a fine predict mask. The coarse to fine workflow is shown in Fig. 3

As shown in Fig. 1, there are two coarse to fine networks C and D. The difference between them is, network C uses the binary mask as the input of fine segmentation network, while network D uses Gaussian smoothed mask as the input of fine segmentation network. The target of using a Gaussian smoothed mask is to reduce the impact of the wrong boundary, but due to our experiments,

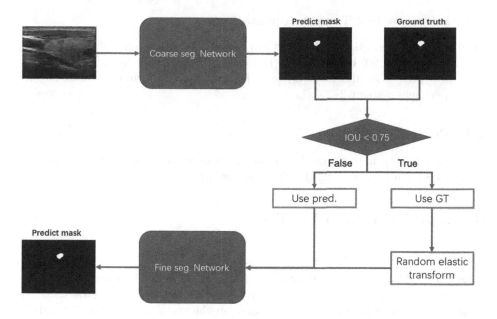

Fig. 3. Coarse to fine work flow

in some cases, binary mask input is better. In contrast, some others with a Gaussian smoothed mask are better, so we use both for the ensemble step.

We tried another coarse to fine workflow. When coarse mask is obtained, a patch will be extracted around the predicted mask, but we have not got a good IOU result in Leaderboard A, so we didn't use this in the final testing for Leaderboard B.

2.2 Loss Function

We use IOU loss as loss function, the initial loss rate is 1e−3, and we also use loss scheduler function 'ReduceLROnPlateau' with 0.6 factor in 10 patience.

We have tried some other loss functions such as BCE loss, boundary loss, dice loss, etc. However, the validation results look very similar.

2.3 Data Preprocess and Augmentation

Data preprocess steps including, image resize to 512 * 512 resolution, intensity normalize to 0–1 range with the float data type.

Data augmentation is significant, especially on the ultrasound image. We test the algorithm based on Deeplabv3+ baseline with and without data augmentation. The IOU metric improves about 9% with augmentation. We use random rotation, random flip, random elastic transformation, and random scale or shift on intensity in our design. Also, random Gaussian noise is applied to the image.

2.4   Inference Test Time Augmentation

Test Time Augmentation (TTA) method is used in our inference step, including original, Rotation 90, Rotation 180, Rotation 270, and their flip images, with a total of 8 predict results for average. TTA improves performance obviously in our experiments, about 2–3%.

2.5   Ensemble

All ensemble step here we used is averaging the predict probability maps. We have tried binary masks but, the result is similar.

3   Experiments and Results

3.1   Dataset

There are 3644 images in training set, during algorithm development stage, 10% cases used for validation and 10%cases used for testing, the other 80% used for training. Table 1 experiments are all based on the training data set. According to these experiments, the IOU metrics result on self-define test set are a little lower than evaluation IOU result with official test data set.

When leadership opened, we separate all training data into 5 cross validation used for coarse network step, inference all predict masks with 5 cross validation models. For other networks, all training data are used in training step. There are 910 images in testing set, all of them are used for testing part.

3.2   Result Evaluation

We tried more than 30 experiments with different baseline, backbone, or parameters during the algorithm development stage. We picked some typical results shown in Table 1.

According to the experiments, Deeplabv3+ has better results than UNet [4] series, PSPNet [5], Hourglass [2], etc. In the table, data aug. means data augmentation including RandRotate90, RandFlip, RandScaleIntensity, RandShiftIntensity, RandGaussianNoise, RandAffine and Rand2DElastic. All of these operations have half probability execution. Because the task is binary segmentation, sigmoid is used here as an activate function.

In post-processing, erode and dilate are also used to remove some outliers.

Some other parameters we set here: batch size = 32 or 64, normalize range = [0,1], image resize to 512 * 512 as input of model.

About the final result, we use 4 individual algorithm branches before ensemble, and their results on leaderboard A are shown in Table 2.

Table 1. Experiment results on self-define test set (365 images)

Baseline	Data process	Loss	IOU
UNet	original data	iou	0.6495
UNet	data aug.	iou	0.7417
UNet++	data aug.	iou	0.6799
U2Net	data aug.	iou	0.7817
shape-attention-unet (densenet)	data aug.	iou+contour	0.7584
PSPNet+(resnet101)	data aug.	iou	0.7796
Hourglass	data aug.	iou	0.7505
Deeplabv3+(resnet101)	data aug.	iou	0.7725
Deeplabv3+(drn)	data aug.	iou	0.7734
Deeplabv3+(xception)	data aug.	iou	0.7861
Deeplabv3+(xception)	data aug.	wbce	0.7820
Deeplabv3+(xception)	data aug.	iou+bce	0.7855
Deeplabv3+(xception)	data aug. + heatmap	iou, smmothL1	0.7732
Deeplabv3+(xception)	data aug. dual channel(img+coarse mask)	iou	**0.7989**
Deeplabv3+(xception)	data aug. + TTA	bce	**0.8025**

Table 2. Experiment results on Leaderboard A

Network	IOU
Deeplabv3+(xception + resnet101 ensemble) + TTA	0.8220
U2Net + TTA	0.8112
Coarse to Fine Net + TTA	0.8079
Gaussian Coarse to Fine Net + TTA	0.8175

4   Conclusion

Our method proposes a coarse to fine two-stage network, which we believe can get good results in ultrasound image segmentation. Also, ensembles are used in the inference step. These improve the final results very much. Data processing is another essential part of the algorithm. Augmentation in training steps with random spatial transform, intensity transform, and TTA are efficient ways to improve results. We tried some methods focusing on nodule boundaries such as boundary loss and attention op rations during the competition, which we think should be helpful. However, the result is not better. We think this can be continued to improve in the future.

References

1. Chen, L.-C., Zhu, Y., Papandreou, G., Schroff, F., Adam, H.: Encoder-decoder with atrous separable convolution for semantic image segmentation. In: Ferrari, V., Hebert, M., Sminchisescu, C., Weiss, Y. (eds.) ECCV 2018. LNCS, vol. 11211, pp. 833–851. Springer, Cham (2018). https://doi.org/10.1007/978-3-030-01234-2_49
2. Newell, A., Yang, K., Deng, J.: Stacked hourglass networks for human pose estimation. CoRR abs/1603.06937 (2016). http://arxiv.org/abs/1603.06937

3. Qin, X., Zhang, Z., Huang, C., Dehghan, M., Zaiane, O., Jagersand, M.: U2-net: going deeper with nested U-structure for salient object detection. Pattern Recogn. **106**, 107404 (2020)
4. Ronneberger, O., Fischer, P., Brox, T.: U-net: convolutional networks for biomedical image segmentation. In: Navab, N., Hornegger, J., Wells, W.M., Frangi, A.F. (eds.) MICCAI 2015. LNCS, vol. 9351, pp. 234–241. Springer, Cham (2015). https://doi.org/10.1007/978-3-319-24574-4_28
5. Zhao, H., Shi, J., Qi, X., Wang, X., Jia, J.: Pyramid scene parsing network. In: Proceedings of the IEEE Conference on Computer Vision and Pattern Recognition, pp. 2881–2890 (2017)

Cascade UNet and CH-UNet for Thyroid Nodule Segmentation and Benign and Malignant Classification

Yiwen Zhang, Haoran Lai, and Wei Yang[✉]

School of Biomedical Engineering, Southern Medical University, Guangzhou 510515, Guangdong, China
weiyanggm@gmail.com

Abstract. The thyroid gland secretes indispensable hormones that are necessary for all the cells in your body to work normally. In order to diagnose and treat thyroid cancer at the earliest stage, it is desired to characterize the nodule accurately. We proposed cascade UNet and CH-UNet to segment thyroid nodules and classify benign and malignant thyroid nodules, respectively. Cascade UNet consists of UNet-I and UNet-II, which segment the nodules in the image at uniform resolution and original resolution, respectively. CH-UNet takes segmentation as an auxiliary task to improve classification performance. We verified our method on the test set of the TNSCUI 2020 Challenge. We achieved 81.73% IoU on segmentation and 0.8551 F1 score on classification, which won the first place in the classification track and was only 0.81% IoU away from the first place in the segmentation track.

Keywords: Thyroid nodule · Segmentation · Classification

1 Introduction

The thyroid gland is a butterfly-shaped endocrine gland that is normally located in the lower front of the neck. It secretes indispensable hormones that are necessary for all the cells in your body to work normally [1]. Until recently, thyroid cancer was the most quickly increasing cancer diagnosis in the United States. It is the most common cancer in women 20 to 34 [2]. In order to diagnose and treat thyroid cancer at the earliest stage, it is desired to characterize the nodule accurately. Thyroid ultrasound is a key tool for thyroid nodule evaluation. It is non-invasive, real-time and radiation-free. However, it is difficult to interpret ultrasound images and recognize the subtle difference between malignant and benign nodules. The diagnosis process is thus time-consuming and heavily depends on the knowledge and the experience of clinicians.[1]

In recent years, deep learning has been widely used in medical image segmentation, classification, detection and other fields. A large number of practices show that deep learning has good performance in medical image segmentation and classification. UNet [3] and ResNet [4] are one of the most popular networks for medical image segmentation

---

[1] https://tn-scui2020.grand-challenge.org/Home/.

© Springer Nature Switzerland AG 2021
N. Shusharina et al. (Eds.): ABCs 2020/L2R 2020/TN-SCUI 2020, LNCS 12587, pp. 129–134, 2021.
https://doi.org/10.1007/978-3-030-71827-5_17

and classification. SE-ResNet is added to the SE-block [5] based on ResNet, and its performance has been improved. In this paper, we use UNet and SE-ResNet to segment thyroid nodules, and classify benign and malignant thyroid nodules.

2 Method

We designed two independent schemes for thyroid nodule segmentation and benign and malignant classification. For segmentation, we proposed a cascade UNet (see Fig. 1). The size of images provided by TNSCUI2020 is inconsistent, the data is first resized to a uniform size and fed to the first UNet. The initial segmentation result of the first stage is used to crop out the ROI region and the ROI is fed to the second UNet. The input image for the second stage keeps the original resolution as much as possible, which can improve the segmentation performance. For classification, we take segmentation as an auxiliary task to improve the classification performance (see Fig. 3).

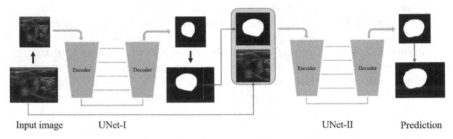

Input image   UNet-I   UNet-II   Prediction

Fig. 1. The pipeline of thyroid nodule segmentation. Two cascaded networks are called UNet-I and UNet-II. All images are resized to 512×512 and input into UNet-I. Then the 512×640 region centered on the initial segmentation result is input into UNet-II.

2.1 Cascade UNet

UNet is the most popular network in medical image segmentation. The encoder-decoder architecture and skip-connection in UNet can capture multi-scale information in medical images. UNet is the preferred network in various medical image segmentation challenges. In TNSCUI2020challenge, the size of the original images and the size of the thyroid nodule vary greatly, so we proposed a Cascade UNet to deal with these issues. UNet-I aims to locate nodules and predict the approximate size and shape of nodules. UNet-II aims to refine the prediction results of UNet-I in order to obtain more accurate nodule boundaries. TNSCUI2020 provides the ultrasound images with a width ranging from 247 to 1280 pixels and a height ranging from 206 to 818 pixels. Firstly, all the images are resized to 512×512 to train UNet-I. There is a trade-off between model performance and computing resource consumption for training UNet-II. Larger input size can keep more information at original resolution, but it also need more GPU memory and time consumption to train the model. Experiments show that 512×640 is considered as the best choice. For images which original size is less than 512×640, we perform padding operation. For images whose ROI in the UNet-I segmentation result is larger than 512 × 640, we resize the ROI to 512 × 640. For the rest images, we crop out the area of 512 × 640 with ROI as the center as input. Therefore, only the image resolution in the second case above is changed. According to statistics, this method can ensure that more than 90% of images keep their original resolution in UNet-II.

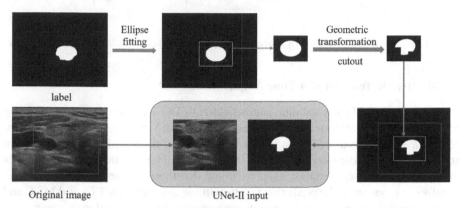

Fig. 2. Flow chart of generating pseudo UNet-I segmentation results. Firstly, the fitting ellipse of the label is generated. The blue box is the bounding box of the ellipse. The green box is expanded 0.3 times on the basis of the blue box. Then do geometric transformation and cutout on the ellipse in the green box. Finally, put the transformed image back to the original image and crop the area of 512 × 640 as the input of UNet-II.

2.1.1 Generating Pseudo UNet-I Segmentation Results from Labels

When training UNet-II, the segmentation result of UNet-I will be used as input. We use labels to generate pseudo UNet-I segmentation results (see Fig. 2). We only hope that UNet-I segmentation results can provide the approximate location and size information of thyroid nodules. It is found through experiments that in the process of training UNet-II, pseudo UNet-I segmentation results should not be too similar to the label, and UNet-II will take a shortcut and copy UNet-I results directly to the output instead of learning how to recognize the original image. Therefore, ellipse fitting, geometric transformation and cutout are used to erase the detailed information in UNet-I results. Geometric transformation includes small scale, translation and 180° random rotation.

2.2 Classification with Auxiliary Task

We take segmentation as an auxiliary task to classify benign and malignant thyroid nodules. Segmentation and classification tasks use the same encoder. The classification head (CH) generates a branch from the bottom of UNet. The classification head includes an adaptive average pooling layer, a dropout layer, and a full connection layer. All images are resized to 512 × 512 and then fed to the network. Our experiments show that the segmentation auxiliary task can improve the classification accuracy. However, the performance improvement of classification to segmentation task is limited.

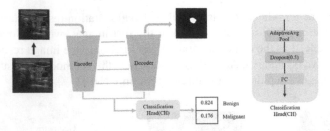

Fig. 3. CH-UNet for classification of benign and malignant thyroid nodules with auxiliary task.

2.3 Multi-scale Test and Test Time Augmentation

Multi-scale test (MST) and test time augmentation (TTA) are common post-processing tricks. MST means transforming the images to multiple different scales in the test phase and then averaging their prediction results. TTA refers to the use of data augmentation and averaging of prediction results in the test phase. They make the prediction results more robust. In the segmentation task, we know that there are targets in each image. Therefore, we can delete the prediction results without any targets in TTA and MST, and only average the remaining results, which can improve the segmentation accuracy.

3   Experiment

3.1   Implementation Details

Training and testing of the network were done in PyTorch[2], all network frameworks are built by segmentation_models_pytorch[3]. The encoders of UNet-I and UNet-II are ResNet34 and ResNet101 respectively, and the encoders of CH-UNet is SE-ResNet50. The network was trained using one Nvidia-RTX 2080Ti GPU. We perform on-the-fly data augmentation including scaling, translation, rotation, flipping, elastic deformation, gamma transformation and Gaussian noise. We performed network and hyperparameter evaluation on initial five-fold cross-validation experiments using the training set of the TNSCUI 2020 challenge[4]. UNet-I and UNet-II are trained using Binary-cross-entropy loss and Dice loss ($0.4 \times \mathcal{L}_{bce} + 0.6 \times \mathcal{L}_{dice}$) and CH-UNet is trained using cross-entropy loss in addition to above loss functions. All networks are trained with Adam optimizer with maximum learning rate 0.0002 (with 1cycle learning rate strategy) and L2 weight regularization factor of 0.0001. UNet-I, UNet-II and CH-UNet are trained respectively with a mini-batch size of 16, 7 and 10 while they are all trained for 300 epochs.

3.2   Ablative Evaluation on Segmentation

In order to verify effectiveness of Cascade UNet, Multi-scale test (MST) and test time augmentation (TTA), we did ablation experiments (see Table 1). We use 384 × 384,

---

[2] https://pytorch.org/.

[3] https://github.com/qubvel/segmentation_models.pytorch.

[4] https://tn-scui2020.grand-challenge.org/.

512×512 and 640×640 as the multi-scale inputs size of TSM. For TTA, we use flipping and gamma transform. Experiments show that both MST and TTA can improve the segmentation performance, and they can improve segmentation performance most when used together.

Table 1. Ablations on segmentation. We validate the effectiveness of Cascade UNet, Multi-scale test (MST) and test time augmentation (TTA) on segmentation task. We show average accuracy (IoU) of five-fold cross-validation on the training set of the TNSCUI 2020 challenge. MST is only used for UNet-I, because the input of UNet-II must keep the original resolution.

UNet-I	UNet-II	MST	TTA	IoU (%)
✓				80.37
✓		✓		80.64
✓			✓	80.59
✓		✓	✓	80.85
✓	✓			81.03
✓	✓	✓		81.20
✓	✓		✓	81.19
✓	✓	✓	✓	**81.34**

Table 2. Ablations on classification. We validate the effectiveness of CH-UNet, Multi-scale test (MST) and test time augmentation (TTA) on classification task. We show average F1 score of five-fold cross-validation on the training set of the TNSCUI 2020 challenge.

SE-ResNet50	Auxiliary task	MST	TTA	F1 score
✓				0.8130
✓		✓		0.8103
✓			✓	0.8100
✓		✓	✓	0.8107
✓	✓			**0.8314**
✓	✓	✓		0.8289
✓	✓		✓	0.8286
✓	✓	✓	✓	0.8296

3.3 Ablative Evaluation on Classification

In order to verify effectiveness of CH-UNet, Multi-scale test (MST) and test time augmentation (TTA), we did ablation experiments (see Table 2). Settings of MST and TTA are consistent with the segmentation task (see Sect. 3.2). The results show that adding segmentation as an auxiliary task can improve the classification accuracy greatly. Experiments show that MST and TTA do not improve the classification accuracy. This may be caused by imperfect setting of MST and TTA. However, due to the limited time, we cannot do a more comprehensive search, and finally we did not do MST and TTA for the classification results.

3.4   TNSCUI 2020 Challenge Results

The TNSCUI 2020 Challenge distributed 3,644 ultrasound images as training set. Subsequently, 400 verification sets and 510 test sets were published. The final score depends on the results on the 510 test set. Our final result is obtained by averaging the 5-fold model results (Table 3).

Table 3.   Challenge results.

Track	IoU (%)	F1 score
Segmentation	81.73%	/
Classification	/	0.8551

4   Summary

We propose cascade UNet and CH-UNet for thyroid nodule segmentation and benign and malignant classification. Cascade UNet can ensure segmentation at the original resolution as much as possible even if the image size in the data set varies widely, which can improve the segmentation accuracy. We find that adding segmentation auxiliary task can improve the classification accuracy, while adding classification auxiliary task does not help to improve the segmentation accuracy. Multi-scale test (MST) and test time augmentation (TTA) are proved to be effective for segmentation, but ineffective for classification.

References

1. https://www.btf-thyroid.org/what-is-thyroid-disorder
2. https://www.thyroid.org/wp-content/uploads/patients/brochures/Nodules_brochure.pdf
3. Ronneberger, O., Fischer, P., Brox, T.: U-Net: convolutional networks for biomedical image segmentation. In: Presented at the International Conference on Medical Image Computing and Computer-Assisted Intervention, Munich, Germany (2015)
4. He, K., Zhang, X., Ren, S., et al.: Deep residual learning for image recognition. In: Presented at the International Conference on the IEEE Conference on Computer Vision and Pattern Recognition, Las Vegas, USA (2016)
5. Hu, J., Shen, L., Sun, G.: Squeeze-and-excitation networks. In: Proceedings of the IEEE Conference on Computer Vision and Pattern Recognition, pp. 7132–7141 (2018)

Identifying Thyroid Nodules in Ultrasound Images Through Segmentation-Guided Discriminative Localization

Jintao Lu[1,2], Xi Ouyang[1], Tianjiao Liu[3], and Dinggang Shen[1(✉)]

[1] Shanghai United Imaging Intelligence Co., Ltd., Shanghai, China
Dinggang.Shen@gmail.com
[2] Department of Control Science and Engineering, Zhejiang University,
Hangzhou, China
[3] Department of Electronic Engineering, Tsinghua University, Beijing, China

Abstract. In this paper, we propose a novel segmentation-guided network for thyroid nodule identification from ultrasound images. Accurate diagnosis of thyroid nodules through ultrasound images is significant for cancer detection at the early stage. Many Computer-Aided Diagnose (CAD) systems for this task ignore the inherent correlation between nodule segmentation task and classification task (i.e. cancer grading). Actually, segmentation results could be used as localization cues of thyroid nodules for facilitating their classifications as benign or malignant. Accordingly, we propose a two-stage thyroid nodule diagnosis method through 1) nodule segmentation and 2) segmentation-guided diagnosis. Specifically, in the segmentation stage, we use an ensemble strategy to integrate segmentations from diverse segmentation networks. Then, in the classification stage, the obtained segmentation result is integrated as additional information along with its corresponding original ultrasound images as the input of the classification network. Meanwhile, the segmentation result is further served as guidance to refine the attention map of the features used for classification. Our method is applied to the TN-SCUI 2020, a MICCAI 2020 Challenge, with the largest set of thyroid nodule ultrasound images according to our knowledge. Our method achieved the 2nd place in its classification challenge.

Keywords: Thyroid nodule · Ultrasound · Identification · Attention

1 Introduction

Thyroid nodule is a common disease with sometimes very rapid growth rates and bad outcomes. Convenient and non-invasive ultrasound has become a commonly used technique for detecting and diagnosing thyroid nodules. However, clinical

---

J. Lu and X. Ouyang—Contributed equally.

N. Shusharina et al. (Eds.): ABCs 2020/L2R 2020/TN-SCUI 2020, LNCS 12587, pp. 135–144, 2021.
https://doi.org/10.1007/978-3-030-71827-5_18

evaluation based on thyroid nodule ultrasound images for distinguishing malignant from benign nodules can be very tedious and challenging. For instance, Fig. 1(a) shows multiple suspicious nodules with varied locations. Their diagnosis process is time-consuming and highly relies on clinicians' proficiency.

To alleviate the burden of clinicians, many Computer-Aided Diagnose (CAD) systems, including advanced deep learning methods, have been developed to identify and diagnose thyroid nodules. One typical type of these methods is to handle the diagnosis task as an instance segmentation task. Many models can be applied to localize and identify nodules [1,2]. Liu et al. [3] extracted diagnosis-oriented features with a multi-branch classification network after a multi-scale detector, while Ponugoti et al. [4] proposed a U-Net [5] based framework for thyroid nodule segmentation with an appended classification branch after the encoder. However, as all these methods extracted the same features for both nodule detection/segmentation and classification, it is not optimal for each task. Besides, different locations in the feature maps may contribute inconsistently to the segmentation and classification tasks. On the other hand, it is known that, by enforcing the model's attention on salient locations, diagnosis/classification can be better achieved. For example, in natural image segmentation, attention to edges of objects could contribute more to segmentation performance. In this way, Song et al. [6] proposed decoupled classification head and detection head after RoI Pooling [7]. However, their framework is limited to natural images; directly application to detection and classification of much subtler and more complex thyroid nodules poses significant difficulty in generating decoupled attention.

In this paper, we utilize the localization cues derived from segmentation results and hereby propose a segmentation-guided attention network to achieve interpretable and accurate diagnosis results. Specifically, an online attention module is introduced to guide the network to focus on discriminative locations with nodules and thus generates a robust diagnosis. We apply our proposed novel network to the TN-SCUI 2020, a MICCAI 2020 Challenge, with the largest set of thyroid nodule ultrasound images to our knowledge, and achieved the 2nd rank in its classification task (F-1 score of 0.8541).

Specifically, we first ensemble different results from diverse segmentation algorithms to refine the segmentation mask by also suppressing false-positive results. Then we feed both the segmentation mask of nodules and the original images into the classification network, trained with an online attention strategy to ensure extracting features from nodule regions for classification, we use an efficient data augment strategy to simulate 1) variations of rotation, position, intensity of nodules in the ultrasound images and 2) variation of patch extraction by randomly cropping patches around the given location of each nodule.

2   Method

In this section, we first introduce the learning-based ensemble strategy for refining the nodule segmentation in Sect. 2.1. A cascaded U-Net (Fig. 2) is applied in the refinement stage after multiple trained networks to reduce segmentation error. Then, in Sect. 2.2, we describe dilation of the refined mask as an input of the classifier while guiding the model's attention onto the segmented nodules through generation and refinement of classification activation map (CAM) [8] in an online manner (Fig. 3).

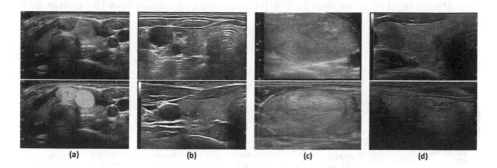

Fig. 1. Examples of diverse thyroid nodule images. (a) A hard case with multiple suspicious nodules (Upper: original image; Bottom: segmentation). (b) Two high contrast images. (c) Two bright intensity images. (d) Two dark intensity images.

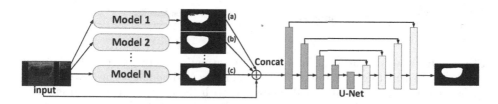

Fig. 2. Ensemble of segmentation results via a cascaded U-Net for generating a refined segmentation mask with precise and smooth nodule margin.

2.1   Cascaded U-Net for Segmentation Ensemble

We apply several high-performance segmentation models (e.g., LinkNet [9] and PSPNet [10]) for thyroid nodule segmentation. Since these semantic segmentation models may over-segment nodules or produce holes inside the segmented nodule (Fig. 2(a)), we add instance segmentation models, e.g., 1) Mask Scoring R-CNN [11] and 2) Deep Snake [2], to constrain the segmentation mask by its bounding box. In particular, the segmentation result from Mask Scoring R-CNN may lead to under-segmentation (lesion outside the predicted bounding box was missed in Fig. 1(a)), which can be compensated by Deep Snake by deforming the boundary of segmentation mask (although Deep Snake may have over-smooth results (Fig. 2(b))).

To take advantage of multiple segmentation models, we propose a segmentation ensemble method via a cascaded U-Net. All the output segmentation masks from multiple segmentation models are concatenated with the original image, and then fed into the U-Net. We use a Dice loss [12] to allow the network to learn from multi-style segmentation masks, thus finally obtaining the refined segmentation mask with suppressed error and also a more precise margin. Note that we can cascade more U-Nets for further improvement, similar to the refinement stages used in Hourglass [13]; alternatively, we can conduct the refinement recursively.

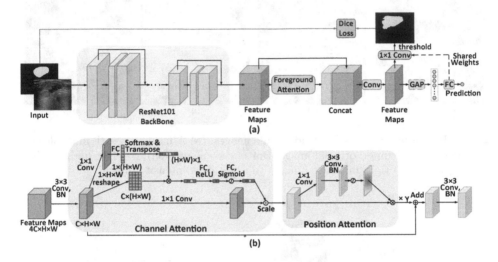

Fig. 3. Overview of our segmentation-guided attention network. (a) The overall pipeline. We extract features from both original image and nodule segmentation mask. Then, we generate CAM online and compare it with the input segmentation mask to guide the locations for feature extraction. (b) Illustration of the foreground attention module. We fuse global spatial information into a channel attention sub-module. Then, the position attention sub-module forwards previous information to the convolution layer for generating the spatial attention.

2.2 Online Attention Module for Classification

We use ResNet101 [14] as a backbone for feature extraction. To keep high resolution feature maps, we change the stride from 2 to 1 in the last residual block; meanwhile, we apply dilated convolution [15] to expand the receptive field.

After the backbone, we add an attention module for both position attention and channel attention. Specifically, inspired by non-local [16] and DANet [17] for global spatial attention, we fuse non-local mechanism into the squeeze-and-excitation [18] attention module by replacing its Global Average Pooling (GAP) [19] layer. Therefore, this sub-module can merge dual attentions in only one step. After that, a lightweight attention sub-module is further applied to enhance the positional attention of our model. We use a 1×1 convolution kernel to reduce the channels, and then apply a 3×3 convolution kernel with subsequent batch normalization and sigmoid activation to get a position attention map.

With CAM [8], the GAP layer can preserve spatial information in the feature maps and enable the classification network to show its attention on the discriminative locations. After propagating weights of the fully-connected layer to the convolutional feature maps, we generate class-specific discriminative regions as the attention map. Inspired by an online mechanism of 3D CAM for COVID-19 diagnosis [20], we leverage an online trainable CAM for this 2D ultrasound image task. Let f represent the feature maps after the last convolutional layer and w represent the weight matrix of the fully-connected layer. As the 1×1 convolutional layer is mathematically the same as fully-connected layer, we copy w as the weight of a single 1×1 convolutional kernel with a ReLU activation function to generate the attention map A as:

$$A = ReLU(conv(f, w)) = ReLU(\sum_{k=1}^{c} w_k f_k), \qquad (1)$$

where k stands for the index of channel with totally c channel, and A is of the same shape of X and Y. Next, we post-process the attention map via upsampling to the original input image size and normalize the values to (0,1). We then add the Attention Loss to maximize the overlap between activation map and the input segmentation mask. Together with the Binary Cross Entropy Loss of classification, we have the total loss:

$$\mathcal{L}_{total} = \mathcal{L}_{Cls} + \lambda \mathcal{L}_{Attention} = BCE(y_{predict}, y_{gt}) + \lambda Dice(CAM, Mask), \quad (2)$$

where λ is the weight factor for attention and is set 0.6 in our experiments. Parameters in the 1×1 convolution layer are always copied from the fully-connected layer, so they will be updated only through the back-propagation of $\mathcal{L}_{Cls}$. $\mathcal{L}_{Attention}$ back-propagates before the GAP layer to guide the feature extractor and the attention module.

After building this mask-guided attention module, we expect the CAM to locate accurately inside or around the nodule segmentation masks, similar to clinicians who diagnose based on some important characteristics on the nodule areas. For instance, a nodule has a higher possibility to be malignant if calcification exists, or a nodule's margin is unclear and hard to distinguish or in a complicated rather than ellipse shape. Also, since all these characteristics indicate aggressive growth, we dilate the ground-truth segmentation mask to include more neighborhood context, which is extremely important for small nodules with very limited internal features. Besides, replenishing neighborhood information from surrounding thyroid tissues is also vital for an accurate diagnosis.

2.3   Data Augmentation

We design an effective data augmentation strategy for robust nodule identification from a large set of ultrasound data with diverse imaging attributes and appearance. As shown in Fig. 1, the majority of images are cross-sectional thyroid tissues with different locations and sizes. For example, an image may include the entire butterfly-shaped thyroid tissue (Fig. 1(a)), or only the left part or the right part (Fig. 1(d)) of thyroid tissue. Hence, we flip images in the horizontal direction (with a 50% chance) to imitate their counterparts. Besides, we randomly zoom out from 0.5× to 1.0× scale for a better view of small objects, and also crop the images covering the nodule areas in a range from the nodule size to the entire image size for simulating various uncertainties. Next, as commonly adopted in the field, we also create rotated and/or shifted images to mimic the real clinical scenario where different perspective images could be acquired when patients are scanned with varied positions and gestures.

Finally, we conduct random image deformations to simulate angles and forces of ultrasound probes. Since this dataset contains images from different ultrasonic machines, it includes different intensity scales and different tissue contrast levels (Fig. 1(b)(c)(d)). To address this, we not only augment image brightness, contrast, Gaussian noise, blurring, and sharpness to enhance the model's robustness, but also normalize intensity distribution of each image to have zero mean and unit variance during pre-processing. In this way, we simulate many possible situations for better approaching real-world application scenarios and minimizing the differences of images acquired from different machines, thus further increasing the robustness of our model.

3　Experiments and Results

The provided training set of the TN-SCUI2020 contains ultrasound images of 1641 benign cases and 2003 malignant cases. To train our model, we conduct a 5-fold cross-validation strategy. Note that, each image is resized to 512×512 as the input to both the nodule segmentation and classification models. We also compare our classifier with other state-of-the-art models using the same cross-validation sets, and the result shows that our model outperforms other comparison methods. Ablation studies also show the effectiveness of each proposed component in our model. To further improve the performance, we finally use majority voting to combine the five trained models from five splits for obtaining robust and accurate results in the official testing dataset with 910 images.

3.1　Ablation Study on Segmentation Ensemble

We compare the refined outputs from the cascaded U-Net with those from the previously trained segmentation models. Using the initial testing split in one cross-validation case with 400 images, 1) LinkNet with ResNet101 backbone achieved 78.99% of mean IoU, 2) PSPNet with ResNet101 backbone achieved 78.50% of mean IoU, 3) Deep Snake with CenterNet detector and DLA34 [21] backbone achieved 76.71% of mean IoU, and 4) Mask Scoring R-CNN with the ResNet50 backbone and FPN [22] achieved 79.28% of mean IoU. Our ensemble method finally obtained 79.68% of mean IoU, which is 0.4% of mean IoU improvement compared to the best of these four segmentation models.

Table 1. Results of nodule classification obtained by six different models.

Model (Undetectable)	ACC	Model (Detectable)	ACC
Fuse-feature	0.7146	Mask Scoring R-CNN (ResNet50 + FPN)	0.7764
ResNet50	0.7523	CenterNet (DLA34)	0.7804
ResNeSt50	0.7712	Segmentation-guided ResNet101 (Ours)	**0.8224**

Table 2. Ablation studies on our segmentation-guided classification model. SG stands for the segmentation guidance for CAM, FA stands for the usage of Foreground attention module. Mask-In indicates that we also feed the segmentation mask into the model. Dil stands for a dilation of 15 pixels to the segmentation mask. We show the mean IoU between the attention areas of CAM and the nodule segmentation masks.

Model	FA	Mask-In	Dil	F-1 Score	Accuracy	Mean IoU
ResNet101				0.7589	0.7500	0.0410
ResNet101	✓			0.7990	0.7896	0.0750
SG-ResNet101				0.8185	0.7910	0.5297
SG-ResNet101	✓			0.8249	0.7923	0.5504
SG-ResNet101	✓	✓		0.8331	0.7964	0.8398
SG-ResNet101	✓		✓	0.8337	0.8005	0.5943
SG-ResNet101	✓	✓	✓	**0.8467**	**0.8224**	**0.9140**

3.2 Performance of Segmentation-Guided Classification

In the experiments shown in Table 1, we compared the accuracy of our model with other CAD models, including 1) Fuse-feature method [23] that combines HOG, LBP, and SIFT features together with the deep features extracted from a fine-tuned VGG Net [24], 2) ResNet50, 3) ResNeSt50 [25], 4) Mask Scoring R-CNN, and 5) CenterNet. For the classification models without a detector, we dilate the ground-truth segmentation masks and crop images around the nodules with context information. Results show that our proposed attention model yielded the best performance and brought an improvement of 4.2% accuracy compared to the best of five comparison models.

3.3 Ablation Study on Classifier

To evaluate the usefulness of each strategy in our classifier, we conducted experiments with different settings. All models were trained with an initial learning rate of 0.00003 and a learning rate decay strategy that decays the rate by a factor of 0.1 every 10 epochs. We set the batch size 20 on 4 NVIDIA TITAN X GPUs and iterated 20 epochs. Backbones were all pre-trained in ImageNet [26].

As shown in Table 2, without the segmentation guidance, the classifier cannot form attention on nodules as its CAM was overlapped poorly with the nodule segmentation mask, while this overlap increased quickly after the guidance of

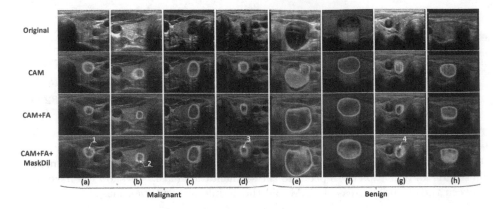

Fig. 4. Generated attention maps for 4 malignant cases and 4 benign cases.

CAM. Experiments also show that combining all our strategies together can achieve a significant boost to the final results, i.e., F-1 score increases and mIoU increases. Note that our model can still generate high accuracy and also locate discriminative nodule areas (Fig. 4) even without feeding any segmentation mask during the testing. These results indicate our model has the potential to be used without requiring the nodule segmentation in the final applications.

We randomly select 4 malignant cases and 4 benign cases, and visualize their attention maps generated by our models in Fig. 4. The second row shows the CAM from our original segmentation-guided attention model. Although the CAM was formed imperfectly with grid shape due to the dilated convolution layers, we can still see their coarse round-shaped attention on the nodule areas. The third row shows the improved results with detailed attention on the nodule areas when we added the foreground attention module. In the last row, we show the improved results when dilating the guided segmentation mask during the training for better attention distribution. Comparing the last two rows, we can see that the attention maps distributed less uniformly and focused more on the key diagnostic areas when dilating the segmentation masks. For example, for the nodules with calcification (e.g., Fig. 4(a)(d)), our model paid more attention on the possible calcification points near the margin of nodules after dilating the segmentation masks (e.g., point 1 in Fig. 4(a)). For the nodules with low and plain inner echo, our model paid more attention on the margin areas with discriminative information, indicating that the model can distinguish the malignant nodules with low margin sharpness and irregular margin shape (Fig. 4(c)) from the benign cystic thyroid nodules (Fig. 4(e)(f)). Instead of always distributing attention to the margins, our model with dilated segmentation masks may shrink its attention to concentrate more on the abnormal components inside a nodule when no useful context can be used (e.g., locations 2,3,4 in Fig. 4(b)(d)(g)).

4 Conclusion

In this study, we proposed a novel segmentation-guided attention network for nodule identification from ultrasound images and achieved a reasonable diagnosis. The use of dilated segmentation masks is able to provide more guidance for the classifier to localize nodules automatically, and finally form more accurate attention to the most informative parts in the nodules for better capturing subtle differences between benign and malignant cases. Experiments show the high performance of our model, indicating also the necessity of providing appropriate guidance to the classifier. Since all testing data were acquired from different machines under different situations, our model has the possibility to be used in real clinical scenarios.

References

1. Cai, Z., Vasconcelos, N.: Cascade r-cnn: Delving into high quality object detection. In: Proceedings of the IEEE conference on computer vision and pattern recognition, pp. 6154–6162 (2018)
2. Peng, S., Jiang, W., Pi, H., Li, X., Bao, H., Zhou, X.: Deep snake for real-time instance segmentation. In: CVPR (2020)
3. Liu, T., et al.: Automated detection and classification of thyroid nodules in ultrasound images using clinical-knowledge-guided convolutional neural networks. Med. Image Anal. **58**, 101555 (2019)
4. Nikhila, P., Nathan, S., Ataide, E.J.G., Illanes, A., Friebe,D., Abbineni, S., et al.: Lightweight residual network for the classification of thyroid nodules. arXiv preprint arXiv:1911.08303 (2019)
5. Ronneberger, O., Fischer, P., Brox, T.: U-Net: convolutional networks for biomedical image segmentation. In: Navab, N., Hornegger, J., Wells, W.M., Frangi, A.F. (eds.) MICCAI 2015. LNCS, vol. 9351, pp. 234–241. Springer, Cham (2015). https://doi.org/10.1007/978-3-319-24574-4_28
6. Song, G., Liu, Y., Wang, X.: Revisiting the sibling head in object detector. In: Proceedings of the IEEE/CVF Conference on Computer Vision and Pattern Recognition, pp. 11563–11572 (2020)
7. Ren, S., He, K., Girshick, R., Sun, J.: Faster R-CNN: towards real-time object detection with region proposal networks. In: Advances in Neural Information Processing Systems, pp. 91–99 (2015)
8. Zhou, B., Khosla, A., Lapedriza, A., Oliva, A., Torralba, A.: Learning deep features for discriminative localization. In: Proceedings of the IEEE Conference on Computer Vision and Pattern Recognition, pp. 2921–2929 (2016)
9. Chaurasia, A., Culurciello, E.: Linknet: Exploiting encoder representations for efficient semantic segmentation. In: 2017 IEEE Visual Communications and Image Processing (VCIP), pp. 1–4. IEEE (2017)
10. Zhao, H., Shi, J., Qi, X., Wang, X., Jia, J.: Pyramid scene parsing network. In: Proceedings of the IEEE Conference on Computer Vision and Pattern Recognition, pp. 2881–2890 (2017)
11. Huang, Z., Huang, L., Gong, Y., Huang, C., Wang, X.: Mask scoring R-CNN. In: Proceedings of the IEEE Conference on Computer Vision and Pattern Recognition, pp. 6409–6418 (2019)

12. Milletari, F., Navab, N., Ahmadi, S.-A.: V-net: fully convolutional neural networks for volumetric medical image segmentation. In: 2016 Fourth International Conference on 3D Vision (3DV), pp. 565–571. IEEE (2016)

13. Newell, A., Yang, K., Deng, J.: Stacked hourglass networks for human pose estimation. In: Leibe, B., Matas, J., Sebe, N., Welling, M. (eds.) ECCV 2016. LNCS, vol. 9912, pp. 483–499. Springer, Cham (2016). https://doi.org/10.1007/978-3-319-46484-8_29

14. He, K., Zhang, X., Ren, S., Sun, J.: Deep residual learning for image recognition. In: Proceedings of the IEEE Conference on Computer Vision and Pattern Recognition, pp. 770–778 (2016)

15. Yu, F., Koltun, V., Funkhouser, T.: Dilated residual networks. In: Proceedings of the IEEE Conference on Computer Vision and Pattern Recognition, pp. 472–480 (2017)

16. Wang, X., Girshick, R., Gupta, A., He, K.: Non-local neural networks In: Proceedings of the IEEE Conference on Computer Vision and Pattern Recognition, pp. 7794–7803 (2018)

17. Fu, J., et al.: Dual attention network for scene segmentation. In: Proceedings of the IEEE Conference on Computer Vision and Pattern Recognition, pp. 3146–3154 (2019)

18. Hu, J., Shen, L., Sun, G.: Squeeze-and-excitation networks. In: Proceedings of the IEEE Conference on Computer Vision and Pattern Recognition, pp. 7132–7141 (2018)

19. Lin, M., Chen, Q., Yan, S.: Network in network. arXiv preprint arXiv:1312.4400 (2013)

20. Ouyang, X., et al.: Dual-sampling attention network for diagnosis of covid-19 from community acquired pneumonia. IEEE Trans. Med. Imaging **39**, 2595–2605 (2020)

21. Yu, F., Wang, D., Shelhamer, E., Darrell, T.: Deep layer aggregation. In: Proceedings of the IEEE Conference on Computer Vision and Pattern Recognition, pp. 2403–2412 (2018)

22. Lin, T.-Y., Dollár, P., Girshick, R., He, K., Hariharan, B., Belongie, S.: Feature pyramid networks for object detection. In: Proceedings of the IEEE Conference on Computer Vision and Pattern Recognition, pp. 2117–2125 (2017)

23. Gao, L., et al.: Computer-aided system for diagnosing thyroid nodules on ultrasound: a comparison with radiologist-based clinical assessments. Head Neck **40**(4), 778–783 (2018)

24. Simonyan, K., Zisserman, A.: Very deep convolutional networks for large-scale image recognition. arXiv preprint arXiv:1409.1556 (2014)

25. Zhang, H., et al.: Resnest: split-attention networks. arXiv preprint arXiv:2004.08955 (2020)

26. Deng, J., Dong, W., Socher, R., Li, L.-J., Li, K., Fei-Fei, L.: Imagenet: a large-scale hierarchical image database. In: 2009 IEEE Conference on Computer Vision and Pattern Recognition, pp. 248–255. IEEE (2009)

Cascaded Networks for Thyroid Nodule Diagnosis from Ultrasound Images

Xueda Shen[1,2], Xi Ouyang[1(✉)], Tianjiao Liu[3], and Dinggang Shen[1(✉)]

[1] Shanghai United Imaging Intelligence Co., Ltd., Shanghai, China
demo.ouyang@gmail.com, Dinggang.Shen@gmail.com
[2] Department of Mathematics, University of Illinois Urbana-Champaign, Champaign, IL, USA
[3] Department of Electronic Engineering, Tsinghua University, Beijing, China

Abstract. Computer-aided diagnostics (CAD) based on deep learning methods have grown to be the most concerned method in recent years due to its safety, efficiency and economy. CAD's function varies from providing second opinion to doctors to establishing a baseline upon which further diagnostics can be conducted [3]. In this paper, we cross-compare different approaches to classify thyroid nodules and finally propose a method that can exploit interaction between segmentation and classification task. In our method, detection and segmentation results are combined to produce class-discriminative clues for boosting classification performance. Our method is applied to TN-SCUI 2020, a MICCAI 2020 challenge and achieved third place in classification task. In this paper, we provide exhaustive empirical evidence to demonstrate the applicability and efficacy of our method.

Keywords: Thyroid nodule · Ultrasound images · Detection · Segmentation · Classification

1 Introduction

Computer-aided diagnostics (CAD), especially in thyroid nodule classification task have a long history. Following huge boost in image classification performance, people started fine-tuning existing networks to classify thyroid nodules [5]. Such fine-tuning, even though exhibits decent performance on certain data sets, could not achieve a universally optimal performance since the fine-tuned networks could be weak in extrapolation. On the other hand, directly fine-tuning networks for classification often misses out entirely on segmentation, which still strains doctors in diagnostics. To address this, there has been an emergence of interest in detection and segmentation methods [4]. On the other hand, there are also questions with regards to whether segmentation should be included since it requires significant computational resources and is relatively easy for radiologists to segment thyroid nodule from image.

© Springer Nature Switzerland AG 2021
N. Shusharina et al. (Eds.): ABCs 2020/L2R 2020/TN-SCUI 2020, LNCS 12587, pp. 145–154, 2021.
https://doi.org/10.1007/978-3-030-71827-5_19

There could be 3 tasks involved for this problem: 1) detection, 2) segmentation, and 3) classification. Ma et al. [1] developed a hybrid model for automatic nodule detection and segmentation. Specifically, it first employs a deep neural network to learn probability maps around ground-truth area. Then, all the probabilities maps are split by the splitting method. Another CNN segments the image from these maps. However, these methods assume a Bernoulli distribution for generating probability maps, which could be questionable. There has been an abundance of literature focusing on thyroid classification. However, most of these methods focus on a somewhat coarse approach by only fitting a pre-trained network to a data set. In the paper by Li et al. [5], they fine-tuned ResNet-50 on a data set featuring different cohorts of ultrasonic thyroid images. Even though they have achieved great performance, this could be partially attributed to the size of their training set with $N = 42952$, while each validation set is only around 1000 images. The disparity in size makes such approach virtually useless in clinical applications. For the classification part, Song et al. [2] created a Multitask Cascade Convolution Neural Network for integrating segmentation and classification. The network features a two-step design. VGG-16 is used as backbone to extract feature maps and recognize nodule coarsely. After this, a spatial pyramid based recognition network finely segments and classifies the nodule. This work integrates segmentation and classification tasks but fails to utilize information generated in segmentation process to aid classification.

In our paper, we answer this question with detailed empirical evidence. Finally, we propose a cascaded network that exploits inherent cues from detection, and segmentation tasks to achieve the final classification prediction that has high sensitivity and specificity, alleviating the workload of doctors.

2   Method

Detection and Segmentation Architecture. Our method aims to complete detection, segmentation and classification. We employed a picture level ensemble strategy by ensembling on masks generated by Mask Scoring R-CNN [9] and CentreNet [11] + Deep Snake [12] combination. In the first branch we employed Mask Scoring R-CNN as a candidate for mask prediction. This network is able to generate high quality masks. However, since the mask generation requires a high threshold, the network produces empty mask for some hard cases in our experiments. To address this issue, we added a two-step segmentation mechanism featuring CenterNet for detection and Deep Snake for segmentation, compensating for lack of detection in Deep Snake. CenterNet [11] is a one stage method for object detection. The network enriches information by centre pooling and cascade corner pooling which mitigates the issue of corner points not capturing image information, thereby increasing detection performance. On the other hand, Deep Snake [12] is an instance segmentation method that is based on circular convolution and contour deformation. It has fast segmentation speed and gives competitive performance. However, due to contour deformation, the mask generated by Deep Snake can sometimes be too smooth. To mitigate inherent drawbacks of both networks, we added a MLP module which consisted of three conv layers for mask selection as shown in Fig. 1.

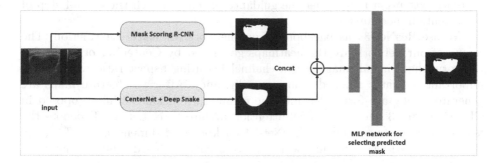

Fig. 1. Demonstration of ensembled segmentation work flow. Two masks are generated by Mask Scoring R-CNN and two-staged segmentation method consisting of CenterNet and Deep Snake

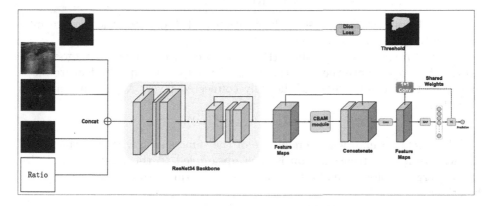

Fig. 2. Two-step attention network. CBAM module is responsible for telling the network where to focus. CAM is responsible for guiding feature maps.

Classification Architecture. Attention mechanism could be used to boost classification performance [13,20]. Class activation map (CAM), proposed by Zhou et al. [13], is able to produce class discriminatory information. The work in its original form, however, allows us to understand the network more but is unable to be directly used to increase the performance since the attention map generated is not integrated in the training process. The method used by Ouyang et al. [14] integrates CAM in an online manner that further boosts the performance of classification network. Convolution Block Attention Module, CBAM, proposed by [17] establishes a method to produce channel features, essentially telling the network "where" to focus. Our network, whose architecture is demonstrated in Fig. 2, utilizes both module. We call this method two-step attention mechanism. After feature maps are generated, the first stage is to produce channel features by CBAM. The online CAM module generates attention map under the guidance of mask, which the second step of our attention mechanism.

We use ResNet-34 as backbone of our network for feature extraction. The inputs of our network are the heatmaps generated by CenterNet of respective category, the original image and a channel featuring aspect ratio which is an important indicator when diagnosing malignancy. After the feature maps are generated, they are forwarded into the channel attention module proposed in CBAM. Figure 3 denotes channel attention architecture. Letting $\mathbf{F}$ denote the feature map generated by our ResNet-34 backbone, of dimension $\mathbb{R}^{c \times w \times h}$, the channel-wise attention module will produce a channel feature map denoted by $\mathbf{M_c}$ of dimension $\mathbb{R}^{c \times 1 \times 1}$. Formally, it is generated by:

$$\mathbf{M_c} = \sigma(\mathbf{F}^c_{max} + \mathbf{F}^c_{avg}),$$

where σ denotes the sigmoid activation. After the generation of this channel-wise attention, we apply this by

$$\mathbf{f} = \mathbf{M_c} \odot \mathbf{F},$$

where $\odot$ denotes element-wise multiplication and f denotes the feature map after applying channel attention.

The feature map and weights of the last fully connected layer undergoes 1×1 convolution to generate the attention map. Formally, let f denote feature maps and w be the weight matrix of the fully connected layer. Attention map A is given by:

$$A = ReLU(conv(f, w)).$$

The attention map will therefore be of the same shape with any channel of the feature map. Attention map is then upsampled to the original input size and undergoes color normalization. We then perform softmasking with sigmoid function:

$$T(A) = \frac{1}{1 + exp(-\alpha(A - B))},$$

where $T(A)$ is the attention map generated by this online attention module. Furthermore, this online module designed a combined loss so that we can calibrate both attention map and our classification results, i.e.,

$$Loss = L_{classification} + \lambda L_{dice}$$

where

$$L_{classification} = BCEloss$$

We use Dice loss to maximize the overlap between attention map and input mask. The classification loss is set to be Binary entropy loss. λ provides a leveraging effect between the two tasks; and since classification is the main task, we set $\lambda = 0.4$.

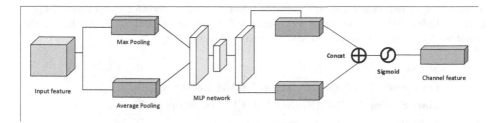

Fig. 3. Channel feature architecture. The input feature are mapped to max pooling and avg pooling separately, then passing through a MLP network.

It should be noted that in training process, the weights of 1×1 convolution layer is an identity map from those of the fully connected layer. The weights of the convolution layer is only updated by $L_{classification}$ since L_{dice} skips the GAP layer in back propagation.

This online, learnable, channel focus CAM module is able to improve the performance of our network and explicitly states area of interest learned by the network. This explainable factor makes the network more interpret-able in its decision making process and would further increase the credibility of the network.

3   Experiment

3.1   Data Set and Augmentation Techniques

TN-SCUI 2020 data set features 3644 ultrasound images of Thyroid gland, of which 2003 are malignant and 1641 are benign. The data set is provided by courtesy of Shanghai Ruijin Hospital. The dataset is then partitioned into training and validation in a 7:3 ratio. To further increase the robustness of our method, we employ a variety of data augmentation methods. Specifically, we randomly rotate the image and apply small degrees of affine transformation to mimic the positions and hardware variances in image acquisition process. Furthermore, we increase the diversity of our data by adjusting brightness, contrast and Gaussian noise. Finally, we train the network on a five-fold, cross validation and cast a majority voting on the testing set featuring 910 images.

3.2   Ablation Study on Ensembled Segmentation

We compare the results of our ensembled segmentation method with those of other models utilized in this paper. The results are shown in Table 1. In particular, we evaluate all of our networks on the testing data set given by the organizers and achieved a 0.3% increase in Mean IoU.

Table 1. Segmentation result of Ensemble UNet. DLA34 stands for Deep Layer Aggregation Model with 34 layer, and FPN stands for Feature Pyramid Network. DLA34 denotes Deep Layer Aggregation Model with 34 layer.

Model	Mean IoU (%)
Deep Snake (CenterNet Detector, DLA34 Backbone)	76.71
Mask Scoring R-CNN (ResNet50 + FPN Backbone)	79.28
Ensembled segmentation (ours)	**79.58**

Table 2. Cross comparison of multiple classification methods. SVM classifier denotes SVM classification on HOG, SIFT and Gabor features. TS-ResNet34 denotes our two-step attention mechanism with ResNet34 as backbone.

Method	Accuracy (%)
ResNet	75.45
VGG	73.82
SVM classifier	72.11
Mask R-CNN	77.12
ResNest50	77.69
CenterNet	77.92
TS-ResNet34	**81.01**

3.3   Ablation Study on Classifier

Table 2 presents classification performance of multiple classification networks. Table 3 presents ablation study on the modules featured in our network and Fig. 4 presents the attention map produced by the network with varying modules applied in the network. We established the superiority of our classifier in the following two regard. First, we demonstrate our method has superior performance to established method in this fielld. On the other hand, we provide empirical evidence suggesting the necessity and edge of our two-step attention module. Furthermore, we have conducted a brief explainability assessment of our network, ensuring our method provides interpretable decision making process to better assist clinical diagnostics.

Combining Table 3 and Fig. 4 gives us a better understanding of our network. Observing images from (2, 1) to (2, 6), we are able to see that without mask guidance, the network was unable to properly guide most of its attention on to the nodule. This situation is best represented by images (2, 4) and (2, 5) as we can that see most of the attention is placed on the perimeter of the image rather than the actual nodule itself. Differences between (2, x) and (3, x), $x \in 1, .., 6$, demonstrate drastic improvements in terms of placement of attention generated by the network. This provides further evidence for inherent correlation between segmentation and classification. Numerically, this is the difference in classification metrics reflected in Table 3, between Serial # b and c, which shows

Table 3. Ablation study of our two-staged attention network. TA stands for two-step, CBAM stands for the usage of the channel attention module, and CAM stands for the usage of CAM attention module. Heat map represents the usage of the class-discriminative detection heat map generated by CenterNet. Ratio represents the addition channel of input consisting of height and width ratio. mIoU represents the mean intersection of union between attention map generated by CAM and segmentation mask.

Serial #	Method	CBAM	CAM	Heat map	Ratio	ACC	F1	mIoU
a	ResNet34					0.7341	0.7477	0.0364
b	TA-ResNet34	✓				0.7240	0.7823	0.0171
c	TA-ResNet34	✓	✓			0.7978	0.8267	0.5016
d	TA-ResNet34	✓		✓	✓	0.7896	0.8188	0.0591
e	TA-ResNet34	✓	✓			0.7814	0.8020	0.5990
f	TA-ResNet34	✓	✓	✓	✓	**0.8019**	**0.8343**	0.5869

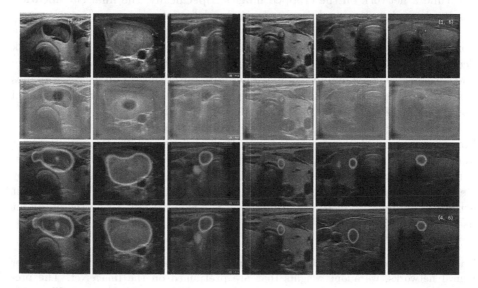

Fig. 4. Instances of the attention map generated by our two-step attention mechanism. Left three are benign cases and the remaining are malignant. First row represents original ultrasound images of the thyroid nodule. Second row represents the attention maps produced by the network without mask guidance. Third row represents the attention map generated with mask guidance without heat map and width height ratio. Fourth row represents the attention maps generated with mask guidance, as well as, with the heat map and the height width ratio as additional inputs. We denote the leftmost image on the first row by (1, 1) and rightmost image at the same row by (1 6). Also, denote the rightmost image in the fourth row by (4, 6).

that the refinement of the attention maps leads to improvements in classification performance. Differences between (3, x) and (4, x), $x \in 1, ..., 6$ represents the difference in attention maps with additional inputs. Even though there is a slight drop in mIoU, the attention regions are more closely fitted to the nodule area, and the rate of change in attention intensity is more continuous. Furthermore, additional inputs are able to diminish opportunities of wrongfully identifying nodules. Looking at differences between (3, 3) and (4, 3), the secondary nodule's attention values are mitigated, which can also be observed between (3, 5) and (4, 5). Such mitigation reduces the risk of misdiagnosis.

3.4   Comparison with Detection-Based Classification

To further illustrate the superiority of the proposed method, we conduct another experiment for this task. We first detect the thyroid nodule with Mask Scoring R-CNN [9] and crop the image according to the proposed bounding box. We then fine tune a network on the cropped images. Specifically, the final classification results depend on the patch images from the detection part.

For detection, we use Mask Scoring R-CNN for proposing the target bounding box [9], which is an improved version of Mask R-CNN [8]. In our method, Mask Scoring R-CNN is not concerned with classification of the nodule, i.e., giving only one category, called Thyroid Nodule. Therefore, after several epochs, s_{cls} comes close to 1, allowing the network to give much of its attention to fine-grained segmentation. Predicted mask is forwarded to Mask IoU head.

For classification, the images are segmented accordingly to the bounding box proposed by Mask Scoring R-CNN. We employed VGG, ResNet, ResNest and Gabor features for comparing classification results. The comparison of classification algorithms are presented by Table 4. The best result is achieved by ResNest, a variation of ResNet employing split attention module [19]. The split attention module produces attention along the channel axis to better highlight useful information for image classification.

Table 4. Comparison of multiple classification methods on the validation set. For neural networks, we adopt weights that are pretrained on the ImageNet. The pretrained networks are then trained on the train set for 30 epochs with an initial learning rate of 0.0001 and undergoing a decay of factor 0.2 every 10 epochs.

Method	Accuracy (%)
ResNet	75.45
VGG	73.82
SVM classifier	72.11
ResNest50	77.69

The above results provide empirical evidence that even though split attention module of ResNest is able to boost classification performance, its performance is still much lower that of our method (i.e., 81.01% in Table 2). This phenomenon demonstrates that, using the localization cues from detection and the segmentation task is suitable for exploring the performance of this task, while roughly cropping the nodule regions may lead to too much misguidance for the final classification.

4   Conclusion

In this paper, we explored inherent connections between segmentation and classification, and designed a two-step attention network to utilize segmentation results for achieving better classification results. Our method achieved the third place in classification at TN-SCUI2020 challenge. Furthermore, our method provides explainable learning by explicitly producing attention maps generated by the network, which we hope would aid doctors in clinical diagnostic process.

References

1. Ma, J., Wu, F., Jiang, T., Zhu, J., Kong, D.: Cascade convolutional neural networks for automatic detection of thyroid nodules in ultrasound images. Med. Phys. **44**(5), 1678–1691 (2017). https://doi.org/10.1002/mp.12134
2. Song, W., et al.: Multitask cascade convolution neural networks for automatic thyroid nodule detection and recognition. IEEE J. Biomed. Health Inform. **23**(3), 1215–1224 (2019). https://doi.org/10.1109/JBHI.2018.2852718
3. Castellino, R.A.: Computer aided detection (CAD): an overview. Cancer Imaging **5**(1), 17–19 (2005). https://doi.org/10.1102/1470-7330.2005.0018. The official publication of the International Cancer Imaging Society
4. Chi, J., Walia, E., Babyn, P., Wang, J., Groot, G., Eramian, M.: Thyroid nodule classification in ultrasound images by fine-tuning deep convolutional neural network. J. Digit. Imaging **30**(4), 477–486 (2017). https://doi.org/10.1007/s10278-017-9997-y
5. Li, X., et al.: Diagnosis of thyroid cancer using deep convolutional neural network models applied to sonographic images: a retrospective, multicohort, diagnostic study. Lancet Oncol. **20**(2), 193–201 (2019). https://doi.org/10.1016/s1470-2045(18)30762-9
6. Prochazka, A., Gulati, S., Holinka, S., Smutek, D.: Classification of thyroid nodules in ultrasound images using direction-independent features extracted by two-threshold binary decomposition. Technol. Cancer Res. Treat. **18** (2019). https://doi.org/10.1177/1533033819830748
7. Guo, D., et al.: Organ at risk segmentation for head and neck cancer using stratified learning and neural architecture search. In: 2020 IEEE/CVF Conference on Computer Vision and Pattern Recognition (CVPR), Seattle, WA, USA, pp. 4222–4231 (2020). https://doi.org/10.1109/CVPR42600.2020.00428
8. He, K., Gkioxari, G., Dollár, P., Girshick, R.: Mask R-CNN. In: 2017 IEEE International Conference on Computer Vision (ICCV), Venice, pp. 2980–2988 (2017). https://doi.org/10.1109/ICCV.2017.322
9. Huang, Z., Huang, L., Gong, Y., Huang, C., Wang, X.: Mask scoring R-CNN. In: 2019 IEEE/CVF Conference on Computer Vision and Pattern Recognition (CVPR), Long Beach, CA, USA, pp. 6402–6411 (2019). https://doi.org/10.1109/CVPR.2019.00657

10. He, K., Zhang, X., Ren, S., Sun, J.: Deep residual learning for image recognition. In: 2016 IEEE Conference on Computer Vision and Pattern Recognition (CVPR), Las Vegas, NV, pp. 770–778 (2016). https://doi.org/10.1109/CVPR.2016.90
11. Duan, K., Bai, S., Xie, L., Qi, H., Huang, Q., Tian, Q.: CenterNet: keypoint triplets for object detection (2019). (cite arxiv:1904.08189Comment: 10 pages (including 2 pages of References), 7 figures, 5 tables)
12. Peng, S., Jiang, W., Pi, H., Li, X., Bao, H., Zhou, X.: Deep snake for real-time instance segmentation. In: 2020 IEEE/CVF Conference on Computer Vision and Pattern Recognition (CVPR), Seattle, WA, USA, pp. 8530–8539 (2020). https://doi.org/10.1109/CVPR42600.2020.00856
13. Zhou, B., Khosla, A., Lapedriza, A., Oliva, A., Torralba, A.: Learning deep features for discriminative localization. In: 2016 IEEE Conference on Computer Vision and Pattern Recognition (CVPR), Las Vegas, NV, pp. 2921–2929 (2016). https://doi.org/10.1109/CVPR.2016.319
14. Ouyang, X., et al.: Dual-sampling attention network for diagnosis of COVID-19 from community acquired pneumonia. IEEE Trans. Med. Imaging $39(8)$, 2595–2605 (2020). https://doi.org/10.1109/TMI.2020.2995508
15. Mayo Clinic: Thyroid nodules (2020)
16. Mayo Clinic: Needle biopsy (2020)
17. Woo, S., Park, J., Lee, J.-Y., Kweon, I.S.: CBAM: convolutional block attention module. In: Ferrari, V., Hebert, M., Sminchisescu, C., Weiss, Y. (eds.) ECCV 2018. LNCS, vol. 11211, pp. 3–19. Springer, Cham (2018). https://doi.org/10.1007/978-3-030-01234-2_1
18. Liu, S., Deng, W.: Very deep convolutional neural network based image classification using small training sample size. In: 2015 3rd IAPR Asian Conference on Pattern Recognition (ACPR), Kuala Lumpur, pp. 730–734 (2015). https://doi.org/10.1109/ACPR.2015.7486599
19. Zhang, H., et al.: ResNeSt: split-attention networks. arXiv:2004.08955
20. Fu, J., et al.: Dual attention network for scene segmentation. In: 2019 IEEE/CVF Conference on Computer Vision and Pattern Recognition (CVPR), Long Beach, CA, USA, pp. 3141–3149 (2019). https://doi.org/10.1109/CVPR.2019.00326

Author Index

Printed in the United States
by Baker & Taylor Publisher Services